T0074976

ALIVE AND WELL AT THE END OF THE DAY

ALIVE AND WELL AT THE END OF THE DAY

The Supervisor's Guide to Managing Safety in Operations

Second Edition

PAUL D. BALMERT

WILEY

Library of Congress Cataloging-in-Publication Data Applied for:
Hardback ISBN: 9781119906650

Cover Design: Wiley
Cover Image: © V. Scott Pignolet/Balmert Consulting

Set in 10/12pt TimesLTStd by Straive, Pondicherry, India

SKY10044715_032023

To All the Good Safety Leaders the World Over

CONTENTS

INTRODUCTION

In the twenty-first century, if you work in industry anywhere in the world you know how important safety is. Whether you own the business, serve as the CEO, are a general manager, department manager or you're a front-line supervisor, managing safety performance—leading people to work safely—is a huge part of your job. When that responsibility isn't carried out effectively, the consequences can be devastating to the people working for the business, the company you work for, and for you, professionally and personally.

You would think that every business that takes safety seriously would teach its new managers and supervisors *how* to manage safety properly. That's often what's done for the other important functions managers and supervisors are responsible for, such as production technology, information technology, accounting, sales, project management, and human resources.

It seems like common sense, but it's hardly common practice. Recognizing that to be the case, shortly after starting up my consulting practice, I developed a practical course on safety leadership. It was called Managing Safety Performance, and its premise was simple: if you're a leader and your job is to send everyone home alive and well at the end of every day, you need to know what to do, and *how* to do that.

That course was written in 2001. In more than two decades since, I've enjoyed the privilege of teaching and consulting with tens of thousands of industrial and business leaders from the front line to the executive suite who work for hundreds of manufacturing and industrial services operations all over the world. In those businesses—from small, privately-held family businesses to the biggest publicly traded industrial names in the world—it's fair that safety is a core value, not something given lip service. Yet a few of these well-managed businesses have a formal process to teach their new leaders what to do and how to do it to achieve the level of safety performance they desire.

Given its importance, you would also think there would be any number of books written on the management practices necessary to achieve great safety performance. Books on great business performance are plentiful, for functions such as sales, marketing, strategy, planning, quality, reliability, customers, and teamwork, written by academics, consultants, and successful business leaders.

You can find numerous books written on leadership, dating back to 1954 when Peter Drucker authored his epic, *The Practice of Management.* Since leadership is leadership, you would assume a book on the topic of leadership and management should be equally applicable to leading and managing safety performance. It might seem that way, but that's not how it works in practice: managing safety involves factors and issues that are unique to safety, and different from everything else in the business. That difference begins and ends with the cost of failure: you can't put a price on human life. Yet there was little to be found in print on the practice of safety leadership in business industry targeted for supervisors and managers working in operations.

Having written and taught courses on the subject, authoring a book was not a giant leap. In 2010, *Alive and Well at the End of the Day* was released by Wiley; the book was intended to fill this void by providing industrial leaders with practical written advice on how to lead and manage safety performance.

The decade that followed produced more than a few surprises.

Alive and Well, as we like to call it, was written with the thought in mind that its audience would be largely based in the US, and primarily consist of front-line leaders looking for practical answers to the question, "What do I do to make sure everyone I supervise works safely?" The second title of the book does read: The Supervisor's Guide to Managing Safety in Operations.

Those two assumptions seemed reasonable: what supervisor wouldn't be looking for help to manage safety performance? But the scope of the audience for those practical answers far surpassed anything that I ever imagined possible.

As things turned out, many readers were leaders well up in the chain of command: Department and Project Managers, Safety and Health staff, Plant Managers and Vice Presidents, and even Company Presidents. In retrospect, I knew leaders at these levels would be much more inclined to read a book about safety leadership than their good leaders serving at the front line.

It would have been better for me to have thought, "This book is written about the front-line leader." But the interest and benefit from the book by senior leaders and executives was more than just to get practical advice as to how to help front-line supervisors lead and manage safety performance better. From firsthand feedback I've received all over the world, it's now clear that leaders at every level have found the principles and practices defined and described in the book to be of value to their practice of leadership.

I should not have been surprised: a number of the safety leadership practices came from leaders I knew who worked well up the management chain of command, including a few Company Presidents and CEOs. On the other hand, most came from leaders whose careers were spent in operations, including front-line supervisors.

The bigger surprise was the degree of interest in the book's content from the rest of the world. *Alive and Well* has been translated into upwards of a dozen languages.

I suppose that should not have come as any great surprise, either. But I will admit that I grew up in a world that was to my thinking limited to forty-eight adjoining states. Now I understand and appreciate how big the world we live really is. Having travelled extensively I can tell you that industrial leaders the world over are united by their desire to see to it that everyone working in their operation does so safely.

Moreover, I can confirm that leaders the world over face virtually the same set of challenges in doing so, simply because people are people. No matter what the industry or the geography, managing safety performance ultimately boils down to leading people to do the right things, and do them in the right way. That being so, why wouldn't every leader in the world be interested in a book describing what works?

As the author of *Alive and Well*, I've been invited to speak to and teach thousands of industrial leaders the world over. Whether speaking in my version of the English language or using interpreters, who are often left puzzling over my use of the language, there has been no finer privilege and pleasure than spending time with these good leaders, sharing principles and best practices to do the most important work of a leader. Friendships have been created all over the world that will last a lifetime. Even better, I've seen ample evidence confirming that these best practices, well executed by leaders, have helped make the world a safer place to work.

Why wouldn't they? The principles and practices came from the best leaders I've seen in action. If the practices made a difference where I worked, there's no reason why they shouldn't make a difference anywhere they are employed.

As gratifying as the experience has been, as I've travelled the industrial world, there are times when I've found myself frustrated with the state of the art of managing safety performance. Leadership practices are highly inconsistent and often subject to the fad of the day; safety performance in some businesses and operations is poor. It shouldn't be that way—and doesn't have to be that way.

The best example is found in the injury rate, which measures the bottom line of safety performance. Within most industries, comparative safety performance is periodically collected and published. The performance differential between the best in their class usually stands in stark contrast with those bringing up the back of the pack. It's not uncommon to find the best twenty times better at managing safety than the worst competitor in *their* industry group. That is exactly what injury frequency rates of .2 versus 4.0 imply.

When that's the case, you would think the worst performers would be studying every move made by their best competitor, asking, "What do they know that we don't?" and "What are they doing that we aren't?" There is without doubt one thing the best aren't doing: spending a lot of time investigating injuries.

Ironically, when it comes to sharing safety processes and practices, seldom is there an unwillingness to share. But it takes two to benchmark, which must start by admitting there is something important to be learned from industrial peers. For some, that is apparently too big a pill to swallow.

Occasionally, in lieu of benchmarking, I'm asked by an executive, "Just tell us what they're doing, so we can follow suit." The question reflects the thinking that benchmarking is a simple matter of copying a successful program. That may work for some things, but most likely won't guarantee improved safety performance.

Why not? As someone who's been inside operations that are world class and those that are farthest from being the best, I can tell you the answer to the question is stunningly simple: the difference that makes the difference is leaders—and their leadership.

As tempting as it is for executives to look to new programs, processes, procedures and standards as the path to improvement and excellence, by my observation and experience safety performance is fundamentally a function of how well leaders do their jobs. Peter Drucker explained it perfectly: "Companies don't compete. Managers compete."

In that sense, safety performance is the purest form of competition between managers. Within any given industry group, the technology is fundamentally the same, as are the safety rules. Companies have essentially the same access to people, and pay is comparable. That means safety performance represents the value added by leaders.

Frankly, that's a message many leaders would rather not hear. They are convinced the fault lies elsewhere, and the way to improvement is to buy a solution. Buying a solution is easy: write a check and install the fix. So that's what they do. When that solution comes up short, they try something else. Improving leadership effectiveness is often the last place executives look for improvement; it should be the first.

As to why it is not, as simple as it seems, improving the quality of leadership is process that demands time, energy, commitment, teaching and coaching. It requires leaders to want to improve themselves, to get better at their leading. Then to be able to coach the leaders they lead. Using coaching as an illustration, to be able to effectively coach up a subordinate in how to lead and manage safety better, the executive or manager serving as the coach must first know what to do, how to do it, and why to do it that way. They must observe performance, analyzing what is being done against what should be done. Finally, they must coach in a way that helps their subordinate perform better. When it comes to the process of safety leadership, there are leaders who have that capability, but I have found them to be relatively rare.

This begins to explain why great safety performance is not the norm. If being great at safety leadership were easy, every operation would be world class at safety.

A second problem with the state of the art of safety management involves its language: conspicuously absent is a common vocabulary for safety. Business leaders the world over know the definition of financial terms such as earnings per share, return on net assets, and gross margin. When I ask what some of the words leaders regularly use in their own practice of managing safety mean, words like risk, culture and accountability, seldom can the leader offer a simple operational definition.

Things get even more interesting when I ask a management team for a common definition; definitions vary significantly. As one example, if leaders don't understand or can't agree as to what "holding someone accountable" means in practical terms, how can their followers be expected to properly manage accountability for them? I've witnessed a heated debate between a CEO and his Safety Director that turned out to be a simple misunderstanding as to the definition of risk, as was used in a newly launched Field Risk Assessment. As the CEO understood it, it was an attempt to undermine what he saw as an inherently risky business; the form was simply intended to identify hazards.

Once, when I asked the management team at a site for their definition of risk, their collective answer was emphatic and uniform: "In our global company, we define risk to include. . ." and they proceeded to rattle off a list of factors such as the severity of harm and the number of people potentially impacted by an event.

While any operation is free to define any term the way they see fit, when the company changes hands or the managers turn over, their definition is subject to change. Which is exactly what happened at this site, when risk was redefined by the new owners.

It's not hard to imagine how these things are seen by those at the front line of the operation. "There they go again." The goal of safety would be better served by a common vocabulary, free of the clutter of the program of the day.

There was a time when quality suffered from the same problem. What is quality? "I know it when I see it." Now because every business is both a customer and supplier, a common definition exists up and down the supply chain. Of course, as any student of industrial history will remind, it took a quality revolution led by thought leaders like Deming, Juran, and Crosby to make that happen. This was not a change coming from the top down but from the outside in.

As to the measurement process, today every industrial operation in the world has become expert in the science and process of measurement. But as sophisticated as operations are when measuring quality, production, and cost, when it comes to measuring safety performance, the principal metrics are little changed from the 70s: injuries and the frequency rate.

That's not to suggest there is anything wrong with measuring what is called in process terms as the output. For safety, that is simply who goes home hurt or hurting at the end of the day. After all, that is safety's bottom line and everyone wants those numbers to be zero.

But, as far as effectively measuring the process inputs that combine to produce that result, or the process as it unfolds, safety metrics are terribly underutilized. Yes, there is interest expressed in what are called Leading Indicators. But the intent of leading indicators is predictive, and rarely is the validity and reliability of any predictive measure called by the name "leading" statistically provable. The safety regulators themselves are guilty of these same failings. And nobody seems to properly understand the meaning of a lagging indicator.

It's one thing to point out these problems and shortcomings. Creating dissatisfaction with the status quo has its place. Actually doing something to make a difference and make things better is a mission I've been on for the last two decades, first by teaching leaders how to lead and manage safety performance, and later writing a book on those safety management principles and practices.

Alive and Well at the End of the Day is my contribution to the list of effective solutions to make the world of work safer. In a sense it is the benchmarking of best leadership practices. It explains the process of leadership as it specifically applies to safety, setting out the means and methods to effectively measure the safety process as it unfolds, in real time and in real life. It also serves to create a common vocabulary of the terms of art that apply to the management and leadership processes.

A reader might ask how I came to this information and knowledge. My process was simple: from more than fifty years of observation and personal experience. *Alive and Well* is not a book on management theory; it's not based on a research study, and it is anything but "the next big thing." *Alive and Well* is simply an explanation of safety leadership practices that work to make a difference, and the principles they are built on.

As to exactly what they are and describing them in a way that is understandable to industrial leaders the world over, that's the one thing I may be uniquely qualified to do. By way of introduction, let me explain what led to what's found in the pages of *Alive and Well*.

As a kid growing up in the 1950s and 1960s, my father, a chemical engineer by education, was employed as a line manager for the DuPont Company. Since you're interested enough to pick up and read a book on managing safety, chances are good you know about the fine reputation DuPont enjoyed for industrial safety: in that era, DuPont was revered as the world's best at industrial safety.

I only knew that Dad headed out the door to work in the morning—or, when he was a Shift Superintendent, the afternoon or evening—and showed up back home, alive and well at the end of every day. But I can tell you I was raised under the DuPont safety culture, not that I understood that until decades later.

Of course, safety culture is what it's called today; back then, it was just how we did things around the house: carefully and safely. I thought that was what everyone did. When we needed some kind of tool to perform a repair, Dad would bring it home from the Tool Room. We drove what must have been the only 1961 Corvair on the planet equipped with seat belts. They weren't original equipment: the belts were installed when Dad brought them home from work one day—an off-the-job safety award from DuPont.

When I turned eighteen, I followed in my father's footsteps into the chemical business. Two weeks after graduating from high school, I got my first real job in the plant where my father worked. It helped that he was the Plant Manager, but not that much: I was hired as a General Helper and assigned to midnight shift, reporting to a Production Foreman, Andy Varab. My first boss.

It was just a summer job, intended to do nothing more than earn spending money for college. The rest of the year, I was a full-time student, attending Cornell University's School of Industrial and Labor Relations. For the four years spent in college, I spent my summer vacation working as a General Helper. Little did I know it would be the beginning of what turned out to be a thirty-year career in the chemical business.

I can honestly say, "I started on the ground floor, on midnight shift. Some days I was sweeping the floor." I had no idea just how valuable the experience would prove to be. At the time, I thought my college classmates getting summer internships in cushy office buildings were the lucky ones. I was fortunate to experience the world of industry from the standpoint of someone who does the real work of the business, at the lowest level in the business. I sat in safety meetings, watched safety videos, was handed safety procedures I was expected to follow, made safety suggestions, and went home tired after working twelve-hour shifts in the summertime heat.

Truth to tell, I had a significant near-miss while driving a forklift; it went unreported. And I got hurt not once, but twice; seriously enough to have my name show up on the plant Injury Report. That did not escape notice by the Plant Manager.

How many eighteen-year old General Helpers experience firsthand what it's like to be held accountable for unacceptable safety performance by a site leader old enough to be their father? It was the first time I got in trouble at work.

"Paul, if that is your approach to doing your job, you are not going to make it to the end of the summer." The conversation took place at the breakfast table at the house on a Saturday morning. Thanks, Dad. It left a lasting impression, not just on my behavior, but my appreciation for the process of managing safety performance.

In four summers, I worked for more than a dozen front-line supervisors; foreman, as they were known back then. Occasionally I will visit a site where that job title still is used; most places these leaders go by Supervisor, Front-line Leader, Team Lead, Coordinator. No matter what the title, the job responsibilities are little changed from when I was working for Andy Varab: production, cost, quality, schedule—and safety.

Still, as college student majoring in Industrial Relations, my lens in looking at my supervisors and their managers was highly unusual: little did they know their management practices were being observed and evaluated by "that kid." Fifty years removed, I can tell you by name who was really good at supervision and who was bringing up the back of the pack, and what made some supervisors far better than their peers.

In their collective support and defense, none were trained in supervisory skills in general, or more specifically, how to get unskilled, untrained (but highly motivated) followers like me to work safely. One day, they were doing the work themselves, quite well to be sure. A job in supervision came open and they were offered a promotion by a leader who'd recognized their work—and their leadership potential. They accepted the job, making the biggest career change anyone in management will ever make: going from doing the work to leading those doing the work, and doing their work safely.

When you think about it that way, being a supervisor is an awesome responsibility.

After college, diploma in hand, I went back to work back at the same plant, this time in a professional capacity. Three years later, I joined what was at the time the second biggest chemical company in the US, Union Carbide Corporation. When I was hired, I was one of 125,000 Carbiders, as we called ourselves, working in operations spread all over the world.

You might associate the name Union Carbide with the Bhopal disaster. In 1984, thousands of people living just outside the fence line of the company's agricultural chemicals plant in India died when a toxic chemical was released. But in the part of Union Carbide where I spent my career, safety was wonderfully well-managed by leaders who I came to know personally. The best of them served as role models for managing safety performance.

I feel as though I was given a gift: spanning three decades, the opportunity to be on a first name basis with more than a thousand leaders; to work at multiple sites including world headquarters; to work with leaders with job titles ranging from front-line Supervisor, Shift Superintendent and Department Manager, Plant Manager and Vice President of Manufacturing, Division Presidents and three CEOs. Add in the Fire Chief and Emergency Director.

Like the different Foreman I'd worked for in my summers working as a General Helper, some leaders were better than others, and yes, there were those who didn't come close to knowing how to lead and manage safety performance. In these ranks were a few top executives with a genius for business, but clueless about safety leadership. I am convinced had they worked for a summer as a General Helper, and reported to a really good front-line supervisor, they would have been far more successful at seeing to it that everyone went home alive and well at the end of the day.

Eventually it became my turn to serve as a line manager. At one point I was responsible for managing more than a thousand people who did the work to operate and maintain a huge chemical plant. I had an advantage: all the things I learned from closely watching the safety leadership practices of the best leaders, as far back in time as those of my first plant manager and the dozen front-line leaders I'd worked for. But it should come as no great surprise that I found out there is no substitute for trying them out on your own.

Training would have helped. I had plenty of good company.

The irony is that while everyone knows how important safety is, rarely are supervisors—or their managers—taught what to do and how to do that to make that happen. On that point, I really wasn't any different than the front-line supervisors I worked for over those summers in college.

The good news is that statistically, serious injuries are relatively rare, and over time, many leaders figure out how to lead and manage safety performance reasonably well. It's what good leaders do. But, even under the best of circumstances, this method of learning comes at a cost to the leader: stress, frustration, mistakes. There are cases where the cost shows up as serious harm to those working for the leader. When that happens, it's been known to leave a lifelong scar on the leader.

Doing something to change that practice was the motivating factor in writing *Alive and Well at the End of the Day*.

Looking back on the thirteen years I spent as a line manager, in the everyday practice of leadership. I now realize my talent wasn't leading: there were others far better at that than I was. But, like a golf teacher who once tried his hand as a competitive player, I turned out to have a different gift: insight.

Not only did I get to see leadership as it was practiced by good leaders from the CEO and executive vice president to the good follower in the machine shop filling in for the boss, I got to experience firsthand the challenges leaders face in sending everyone I was responsible for home alive and well at the end of the day. I was fortunate that most days this was successfully accomplished. But there were days I wound up visiting my followers and their family and friends at the hospital.

Now, more than a decade removed from its release, this is the perfect time to revisit what was written and to create a second edition of *Alive and Well*. It's not because the book's principles and practices have become outdated; nothing could be further from the truth. To have found their way into the first edition, every principle and practice must have stood the test of time and use in the real world of operations.

Yes, the times change. People change and so does the culture. But some things do not: there will always be people who do the work of the business, and hazards present as they do their work capable of hurting them. As desirous as anyone might be to work safely, leadership will always be required to make that happen.

As to the practices of effective safety leadership, those based on the principle of treating people the way the leader would want to be treated will never have to change with the times.

The teachers in our practice and I have another decade of experience teaching them to industrial leaders the world over. In the process, we've gained more insight, discovered more about the process of leading and managing safety, and learned how to explain things better. We realize there are principles that deserve a more complete and detailed explanation: the principle of Performance Visibility is the best example. Sadly, significant tragic events have occurred around the globe that offer more lessons that serve to illustrate the book's principles and practices and underscore their value.

I am confident what you will find in *Alive and Well* will help you protect the people you supervise on the job. I can think of no higher business or professional objective any leader can have than to do exactly that.

Paul D. Balmert

ACKNOWLEDGMENTS

Alive and Well describes the successful practice of safety leadership in the real world and real time of operations. The principal source of the book's content comes about from a thirty-year career in the chemical industry, where I worked for and around more than a thousand leaders who I knew on a first name basis. Since starting up a consulting practice in 2000 it's been my privilege and pleasure spending quality time in the classroom and conference room with upwards of a hundred thousand industrial leaders the world over. As the saying goes, "he who teaches learns the most." There are my good colleagues and friends in our consulting practice, who themselves had successful careers: they, too, have given me so many good ideas about the practice of safety leadership.

This book isn't so much by me as it is through me. My contribution to the world of safety leadership comes mainly by simply paying attention.

It would be impossible to recognize the contribution made by each and every one of these fine leader's whose ideas and practices are reflected in the book. There certainly are cases where I distinctly recall the circumstances and exactly what was said and done. But in so many other cases, I was part of a safety culture that had a positive influence on me in subtle but very powerful ways. Leaders deserve full credit for that.

The content of this book reflects what I learned from all these good leaders. For their contribution to *Alive and Well*, individually and collectively, I thank each and every one. They've made a huge difference in my life.

PDB

ABOUT THE AUTHOR

Since its establishment in 2000, Paul D. Balmert and his colleagues at Balmert Consulting have helped improve safety leadership practices and safety performance for clients throughout the world and have been active in a wide range of industries including mining, manufacturing, maintenance and construction, oil and gas exploration, production, refining and distribution, chemical manufacturing, the paper industry, and power generation. Paul and his associates have taught safety leadership principles and practices to nearly 100,000 industrial leaders, from front-line supervisors and safety committee members to company presidents, CEOs and board members. A highly effective communicator, Paul has spoken frequently in seminars and conferences. For two decades, he has published a newsletter, *Managing Safety Performance News*, that's widely read and reprinted in industrial safety circles.

Prior to establishing Balmert Consulting, Paul acquired valuable experience and insight into the process of managing safety performance in a 30-year career in chemical manufacturing, primarily with Union Carbide Corporation, in a variety of roles at the site, division and corporate office. He served as a line manager for 13 years, responsible for managing maintenance, shift production and distribution operations at the company's largest manufacturing facility. In his career he also managed business support functions including human resources, operations training, emergency response, and public relations. Paul is a graduate of Cornell University's School of Industrial and Labor Relations. His practice is based in Seabrook, Texas, near Houston.

CHAPTER 1

THE LEADERSHIP CHALLENGE

> Leadership commitment without capability won't produce excellence.
>
> —*Plant Manager Van Long*

What do I do to lead people to work safely? How do I manage safety performance? These are two of the most important questions every leader in business and industry needs to ask. But, asking is not enough: every leader in business and industry must have practical answers to these questions that can be counted on to work in the real-life world of operations.

That is exactly why *Alive and Well at the End of the Day* was written. It's a book about the practice of safety leadership, written for leaders in business and industry the world over.

CALLING ALL LEADERS

Who are the leaders in business and industry? What makes them leaders? In a book written on safety leadership, answering these two questions is the perfect place to begin.

Likely you're a leader. You may hold a formal position of leadership: supervisor, manager, or executive. Normally, when you hear the term supervisor, it's used to describe the front-line supervisor. However, the definition of a supervisor is someone responsible for the work of others; in that sense the CEO and the plant manager are supervisors, too.

Yet, in practice, not every leader is a supervisor—and not every supervisor is a leader. There are always leaders who don't hold formal positions in the ranks of management and supervision. There are leaders in staff organizations: safety leaders can usually be found in the Safety Department. There are safety committee members, safety stewards, and on every crew, peer leaders, who for better or worse, have their influence. Technical leaders are looked up to when troubleshooting an operating problem. Have an emergency, everyone takes orders from the Fire Chief.

Alive and Well at the End of the Day: The Supervisor's Guide to Managing Safety in Operations, Second Edition. Paul D. Balmert.
© 2023 John Wiley & Sons, Inc. Published 2023 by John Wiley & Sons, Inc.

Collectively, industrial leaders are an incredibly diverse lot. Many are appointed and promoted; some are recognized by virtue of their technical know-how; others are situational leaders, like the pipefitter who serves as the Fire Chief when the emergency alarm sounds. At the crew and individual contributor level, there are natural leaders: people seem to be naturally attracted to following them.

Study them carefully, you'd be hard-pressed to find common core competencies, even among the best at leading. Great leaders can be extroverts or introverts; caring and feeling or unemotional and distant. Some leaders have technical expertise and the wisdom of experience; others have street-smarts and the exuberance of youth. There are great communicators and brilliant strategist; boring speakers and those amazingly gifted at getting things done.

Leave it to the noted management consultant, Peter Drucker, to see what may well be the only common denominator across the spectrum of leadership strengths and styles: followers. "Leaders have followers," he wrote. His three simple words—leaders have followers—reveal much about the process of leadership (Figure 1.1).

Without giving the subject of leaders and leadership much thought—let alone thoughtful analysis—it's easy to think of leadership as being all about the leader, and to look upon followers as having little say in the process. But in a very real way, followers hold huge power, as it is the followers who make the leader. Yes it is true that in business, a supervisor or manager will be promoted and given a leadership role, and yes, subordinates will follow the supervisor simply because they know what can happen if they do not. But in the practice of leadership, there's a world of difference between grudging compliance and true followership.

Followers are the ones who make that choice.

That explains why, in what we will call the Leadership Model, the arrow connecting leaders and followers is drawn in the upward direction: *from* the followers *to* the leader. Leadership for any function or goal is a process of getting followers to follow.

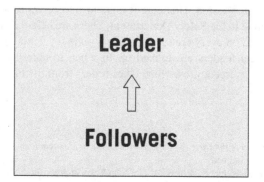

Figure 1.1   Followers should be looked upon as equal partners in the leadership process.

Understood in that way, leadership can be appreciated as a game played in the five and half inches between the *follower's* ears. That view begins to explain why leadership is such a great challenge—it's about creating followership—and why so many attempts to lead so fall short of the mark: followers simply don't want to follow.

In many cases, who could blame them?

THE SAFETY LEADERSHIP CHALLENGE

Running a business producing products and services makes leaders in operations responsible for production, cost, quality, schedule, reliability, business process improvement, morale, teamwork, customer satisfaction, environmental performance and *safety*. Making all that happen the way it's supposed to is a daunting challenge: every one of those business functions is important and each function comes with a nearly unlimited supply of things that can go wrong. They often do. That's why leaders in operations are some of the busiest people on the planet.

In theory, of all of those responsibilities, safety should by far and away be the easiest to manage. The goal of safety is simple: seeing that every follower goes home alive and well at the end of the day. Either they do or do not.

If you're a front-line supervisor, working for a large industrial operation, you're part of a management team. You've got a manager, a site leader, a director of operations, a vice president of production, a division president, a company president, and a CEO. Do all of those leaders above you want to see that every one of your followers goes home safe? Of course they all do.

What about your followers? If you asked, "Do you want to go home safe at the end of the day?"—what would they tell you? "Of course we do. Nobody wants to get hurt." The answer seems so obvious that leaders seldom bother to ask the question. "It goes without saying." Leaders should ask: it makes a point.

But the point served here is to demonstrate a fundamental truth about safety: perfect goal alignment exists from the top of the outfit to the bottom. Everyone wants to be safe. Every follower wants to work safely; every leader wants every follower to work safely. This alignment is not a matter of people nodding in agreement when there is none; it's not a case of checking the box saying there is alignment. Rarely can that be honestly said for any other business objective.

That degree of true alignment *should* make managing safety easy: the easiest of all the responsibilities a leader has.

That's the theory. In practice, managing safety performance is a brutally tough challenge. It might be easy to send everyone home alive and well at the end of a single day; making that happen day after day, week after week, month after month, year after year is daunting challenge. Being safe yesterday guarantees nothing today.

What is standing in the way? What makes the challenge of safety that tough?

That question has been asked to a hundred thousand leaders the world over—leaders from CEOs and board members, to front-line supervisors and informal leaders; leaders in business and industries such as agricultural products, construction, chemicals, energy exploration, production, refining, manufacturing, maintenance, pulp and paper, mining, steel fabrication, transportation, utilities; leaders for businesses spanning the globe; leaders in small, family-owned operations; leaders who speak dozens of different languages.

No matter who the leader, what the business, where the location, or what the language spoken, the answers to the question are virtually identical and totally predictable.

In practice, leaders face a long list of tough safety leadership challenges; arranged in alphabetical order, they look like this:

1. Attitude: getting people to have a positive view about safety
2. Awareness: getting people to recognize hazards
3. Behavior: changing behavior so people work safely
4. Change: buy into change instead of resisting
5. Complacency: convincing people that they can get hurt doing familiar work
6. Compliance: getting people to follow all the rules—all the time
7. Communication: finding out what's really going on
8. Contractors: depending on third parties to work safely
9. Culture: changing the common practices everyone does
10. Distractions: keeping people focused on the task at hand
11. Environment: work is done in a tough physical environment
12. Equipment: working around hazardous equipment and fixing defective equipment
13. Getting people to... (fill in the blank)... wear PPE, stop unsafe jobs, report near-misses
14. Pressure: from peers, customers, schedule, and leaders
15. Production versus safety: taking care of the business and taking care of safety
16. Risk-taking: dealing with the risk-takers.
17. Safety meetings: running safety meetings that aren't a waste of time
18. Training: getting new people trained; providing training that creates learning
19. Turnover: replacing skills and knowledge lost
20. Zero: with all those challenges, how to achieve zero harm?

Are you surprised by anything on the list? Likely not. Take the time to identify all the challenges you face in managing safety, you'll come up with a similar list. You would do well to do exactly that.

As long as this list is, and as tough these challenges are, the biggest single challenge leaders face the world over isn't even on this list!

But leaders are too busy dealing with their challenges to spend any time thinking about them. Or doing the research, benchmarking, and observation of the best practices needed to figure out the "what to do" and "how to do it" to deal effectively with these real-world leadership challenges.

As to what gives rise to these challenges, by and large the common thread is people. Essentially these challenges come about because people work in an environment where there are hazards, and everyone knows why they are there: to get the work done.

To provide *practical* answers to these kinds of challenges is exactly why this book was written, and what it is intended to do. Sure, there's a place for academic research and leadership theory about safety management. And yes, in theory, engineering out every hazard would solve every safety challenge. Then it wouldn't matter what people do, because they just couldn't get hurt. But that's simply not practical.

Leaders must have practical solutions guaranteed to work. They need these solutions in a readily usable form: simple, practical, and effective.

If you're facing the same kind of leadership challenges as do your peers all over the world—attitude, awareness, behavior, compliance, complacency, culture... zero—this is the right book for you. The answers are here: proven techniques successful leaders, in operations all over the world who have faced the same challenges, use to ensure that people work safely.

This is a book written for use by leaders in real life—in real time.

You'll find the answers laid out, one challenge at a time, one chapter at a time, in a logical sequence. For many of the challenges there are real-world scenarios to illustrate how the challenge manifests itself out on the shop floor, and examples of how the techniques might look when properly executed. You won't need much imagination to picture the problems—most likely you'll think the situations came straight from your operation, maybe even from one of your followers.

LEADERS MUST LEAD!

These challenges are tough, universal and totally predictable. They don't show up all at once, but over time leaders will face almost all of them. For example, the senior, experienced followers easily get complacent; when they are promoted or retire, their replacements lack training and experience. Eventually the new people will become complacent.

But have the unfortunate experience of a major incident, the next day nobody will be complacent. It's as if facing complacency is the reward for succeeding in getting followers to work safely.

So, what do you do about your tough safety challenges?

It's not enough to make the list; if nothing is done to solve the challenges found on the list, sooner or later someone will go home hurt. If you doubt that, look at the root causes for the injuries that happened in your operation and see how they compare with what's on your list. When you read findings like "failure to recognize the hazard," "rule known but not followed," "common practice," "in a hurry," "defective equipment," you're getting confirmation that you've identified the right challenges.

Not that you need it. Most leaders know exactly what they're up against.

You could hope the challenges will take care of themselves. But hope is not a method. It would be nice to think your followers would solve the challenges for you: after all, it is *their* safety at risk. The great paradox of safety is this: all followers want to go home alive and well at the end of the day, but without strong and effective leadership, that simply will not happen. Your boss might decide to intervene on your behalf; but if that's what it takes, your supervisor would have to question why you're even needed.

Bottom line: there really is no escaping the fact that no matter what your job title, if you are a supervisor—responsible for the work of others—managing safety performance is *your* job.

Leaders must lead!

The injury rate an operation experiences reveals how well the leaders *collectively* perform their job of taking on these tough safety challenges. The injury rate reflects the collective value created by the leaders to the business's safety performance. A comparison of the injury rates across any industry is bound to show a huge difference between how well the best do, compared with the rest of their industrial peers.

That difference is found in their safety leadership practices.

THE FRONT-LINE SUPERVISOR

No leader plays a more important role or contributes more value to the bottom line of safety than does the front-line supervisor. Not that they typically see it that way. Ask a front-line supervisor, "Who has the real power in your organization?" they invariably point in the direction of the front office. "That's where the important decisions get made." The widely held perception is that those few at the top of the organization wield the power and make all the difference.

It's true that the role of middle managers and front-line supervisors is to get behind their leaders and carry out what's expected. The organization pyramid neatly sums up that view. It may well be the oldest management concept on the planet, one that everyone understands all too well. The higher up in the pyramid you move, the fewer leaders there are, the more important those leaders are, and the more power those leaders have. Follow the model to its logical conclusion, and you find the person at the pinnacle of the pyramid. In the conventional wisdom of organization power, that's the most important and powerful person in the organization.

There is an alternative way to view the relative value and importance of those working for a business. It begins with the fundamental purpose of the business, why the business exists in the marketplace. In private enterprise the goal of a business is ultimately to create value for the owners. It's the *doing* of the work that creates economic value for the owners. That value-creating work is not a management function.

CONVENTIONAL WISDOM

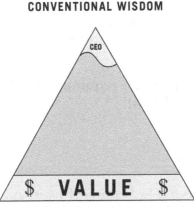

Figure 1.2 The conventional wisdom about value in the hierarchy is wrong.

ECONOMIC REALITY

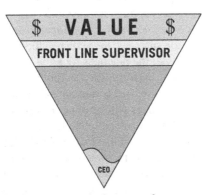

Figure 1.3 Front-line leaders represent the interests of management—right up to the CEO—at the point where economic value is created.

Thus, while conventional wisdom suggests one view of hierarchy, economic valuation suggests and opposing view (Figure 1.2).

The Inverted Pyramid (Figure 1.3) presents the concept of who really matters in an entirely different light: those who perform the work and create value really are the most important people. As to management, its role is helping those who create value, do their jobs well and better. More than a few smart leaders figured that out a long time ago.

Take that economic logic a step further, and there is a second very interesting and significant conclusion to be drawn. Of all who work in the ranks of middle management, the Inverted Pyramid suggests the front-line supervisor is next in line in importance. In business, the front line of supervision is the point of execution: where the product gets made or the service is delivered. The job of the front-line leader is to represent the interests of all the other members of management right up to the top, seeing to it that this vital work of the business is done, done well, and done safely.

In practice, it's the front-line leaders who have the most direct impact on what goes on, and the most complete understanding of what is going on. In that sense, they truly are the most important level of management in the enterprise.

This isn't just inverting the pyramid: it's standing conventional wisdom on its head!

READING THE BOOK

No matter how high the level of leadership commitment to safety, the capability of the leaders determines the outcome. Effective leadership requires motivation *and* technique. That's where *Alive and Well* fits in to the safety leadership process: it defines and teaches techniques in the form of leadership best practices. It teaches more than *what* and *how;* it explains why to do it that way, and *how* it works. In that way, *Alive and Well* creates understanding, not just knowledge.

The subtitle for *Alive and Well* reads, *The Supervisor's Guide to Managing Safety in Operations.* As the title suggests, the audience for the book is first and foremost the front-line leader; many of the practices found in the chapters are directed at the specific leadership duties they perform, such as running safety meetings, managing safety suggestions, and observing and correcting behavior.

That does not mean the practices will be of no interest to managers and executives. On the contrary: leadership is leadership, no matter what the level or who the followers. All are best practices. Many of these safety leadership practices are appropriate for every leader at every level; in some cases, the practices came from the observation of top executives in action. Moreover, when an executive practices a best practice, the influence of their role-modeling on their followers cannot be overstated.

There is also the important leadership function of coaching up followers who are themselves leaders on their management practices. Coaching isn't the same thing as teaching or counseling; it's more akin to what a good sports coach does to help an athlete perform better. The elements of coaching are observation, analysis, and providing useful help. At the analysis step, the best practices described in *Alive and Well* are invaluable: they establish the model of performance against which what's observed can be compared. Practices describe, in a very practical and detailed way, exactly what "good" looks like. Without that model, it's guessing, not coaching.

Most of us grew up being taught to read books from beginning to end. The presumption is that a book is written sequentially, each chapter building on the one preceding. By that logic, you can't understand and appreciate the last chapter unless you've read all the chapters leading up to it.

Of course, the presumption made by the author is that the reader has ample spare time to read the book from beginning to end. In this day and age, for leaders in operations, that's not realistic.

Alive and Well at the End of the Day started out to be a short book—the kind of book a busy leader like you might actually have the time to read, cover to cover. There was a problem with that: the challenges leaders face in managing safety performance are too numerous and too complex to be dealt with in a brief or incomplete way. So, to do the subject of leading safety performance justice, *Alive and Well* has become a relatively long book. You would do well to read it from beginning to end—if you have the time.

But you don't. Therein lies a very fundamental problem. Your time as a leader is precious. You're looking for help, but don't really have the time to wade through pages of written material to find it.

That's why you'll find **The Guide** in this opening chapter. Think of it as the troubleshooting guide in the owner's manual of your vehicle. You know how that works: a condition, symptom, or problem is described, followed by instructions for fixing the problem. It's a great approach to getting the information you need—fast. No need to spend time on what's not a problem. At least until it is.

This book wasn't written on the assumption that it must be read from beginning to end, or that you have the time to do that. If you don't, here's the alternative approach: read the chapters that offer immediate help for the specific challenges you face right now. Save the other chapters for later, when need dictates or time permits.

That said, if you go to the chapter on your toughest safety challenge, it's likely you'll find the recommended practice incorporates, a second tool explained earlier in the book. For example, if you're looking for immediate improvement in the effectiveness of your safety meetings, you'll find the recipe for success in Chapter 14, "Safety Meetings Worth Having." But the solution uses a technique explained in Chapter 9, "The Power of Questions."

Still reading two chapters to improve safety meetings is a lot easier than reading an entire book.

THE CASE FOR SAFETY

No matter what the challenges you face, Chapter 2, "The Case for Safety," should be the place to start. "The Case for Safety" describes the *real* reasons why safety is always the most important thing for every leader in operations to get right. More important than getting the product to the customer or making sure the customer is satisfied with the work.

Even if you're already sure you know the reasons, it's still a chapter worth reading. It explains the most fundamental idea in the book. But then, the basics never go out of style, and, sadly, there have been more than a few leaders who never fully appreciated "The Case for Safety" until it was too late. You'll read some of their stories in this chapter and will be able to appreciate why you never want to be in their shoes!

Once you've read Chapter 2, go for solutions to your most pressing problems. In *Alive and Well* you'll find ways to deal with each of the specific safety leadership challenges in the chapter list. In each chapter you'll find a fuller explanation of the nature of each of these challenges: why each isn't some minor irritant and can't be easily solved by a safety meeting or another safety policy, and why leadership is required to make a real difference. Often the challenge will be illustrated by a case study or scenario.

Then we'll delve into potential solutions. Instead of theory, we'll offer concrete strategies and tactics—what to do and how to do it—that have been successfully used by leaders. In each chapter we'll also show you, with examples, how to implement these techniques. See Figure 1.4 for **The Guide** to this book.

THE GUIDE

Accountability	When things go wrong, people aren't held accountable.	Chapter 12
Attitude	How do I change people's attitude to get them to work safely?	Chapter 8
Awareness	People don't see the hazards that can harm them.	Chapter 7
Change	I'm constantly faced with changing safety policies and procedures and face resistance.	Chapter 10
Changing Behavior	How to I change bad habits?	Chapter 8
Compliance	How to I get people to follow all the safety rules we have?	Chapter 6
Complacency	How to prevent my crew from becoming complacent.	Chapter 7
Control	As the leader, I may be responsible but I have no control over what happens.	Chapter 19
Culture	How do I create a safety culture in my organization?	Chapter 15
Equipment	I'm handcuffed by the equipment I've been given to operate, but there isn't a lot of money available for upgrade.	Chapter 19
Experience	A lot of our people are new and don't have a great deal of experience.	Chapter 16
Execution	If only I could just get the people in my organization to do all the things they are already supposed to do.	Chapter 21
Influence	As just one leader in a big organization, can I really do anything that will make a difference?	Chapter 22
Investigations	How do I get people to own up to what actually went wrong – and what they did to contribute to the situation?	Chapter 11 Chapter 12
Leading	What do I actually do as a leader to get people to work safely?	Chapter 3

Figure 1.4   The Guide.

Lessons Learned	What can I learn from the mistakes my peers have made?	➡	Chapter 20
Measurement	What kind of leading indicators can I use to tell me about safety performance?	➡	Chapter 17
Management	If my leaders don't change, we'll never be able to achieve great safety performance.	➡	Chapter 15 Chapter 21
Near-misses	I know we have many near-misses that I never hear about.	➡	Chapter 7 Chapter 11
Positive Reinforcement	Do I need to give compliments?	➡	Chapter 8
Production vs Safety	We have a business to run and customers to serve. How does safety fit with that reality?	➡	Chapter 2
Risk	Can we run our operation with zero risk?	➡	Chapter 7
Risk Taking	My people are taking risks that I think are unacceptable.	➡	Chapter 7 Chapter 8
Safety Meetings	Our safety meetings are just plain boring.	➡	Chapter 14
Safety Suggestions	I can't remember the last time I got a good safety suggestion from someone on my crew.	➡	Chapter 13
Solutions	When there has been an accident, how can I come up with a solution that really will prevent it from happening again?	➡	Chapter 11
Surprises	I keep getting surprised by safety problems.	➡	Chapter 11 Chapter 21
System	We have a robust system to manage safety performance. Can I count on that to keep people safe?	➡	Chapter 7 Chapter 18
Time	I have so many things on my plate that I just don't have enough time to manage safety performance.	➡	Chapter 4
Training	How do we make sure our training courses are up to the test of teaching people what they need to know to work safely?	➡	Chapter 15

Figure 1.4   Continued

All that is designed to help you improve safety performance, or, if your performance is as good as it gets, maintain the level of excellence you are currently achieving.

Getting there is half the battle; staying there is the other part.

JUST IN CASE

You might be one of the lucky few in the ranks of management who have managing safety performance down to a science and don't see the need for any help. If so, luck probably has little to do with your success. But if you have any nagging doubts about whether you really do have all the bases covered, try reading Chapter 20, "Mistakes Managers Make." If nothing else, that chapter will provide a cross-check to ensure you haven't overlooked something that might set you up to fail.

CHAPTER **2**

THE CASE FOR SAFETY

> Knowing something—and understanding it—are not the same thing.
>
> *—Charles Kettering*

Alive and Well is a practical book written for safety leaders in industry and business the world over. It's intended to answer the question every leader should ask: to lead safety well, *what* do I do, and *how* do I do that? That we will do, but after we first answer the single most important question every leader must ask: *Why* is safety the most important goal of every leader in business?

In the twenty-first-century business world it's just about impossible to find a leader from the CEO down to a newly promoted front-line supervisor who doesn't know *that* safety is important. Everybody knows that, but *that* isn't the question we're looking for the answer. The question is *why*.

The answer to the question "Why is safety that important?" seems rather obvious. No leader would ever want to see someone working under his or her supervision get hurt. If that's not enough, leaders have been told by their leaders, right up to the top of the company, that safety is a key part of their job duties and performance measures. Getting good safety performance is a requirement for a successful career.

Moreover, safety is no longer just "important". In many organizations, safety is seen as a core value. You can find that in a mission statement and on a list of company values posted on the walls of conference rooms. You might think that should be more than enough to settle the issue once and for all. Safety really is that important. Case closed. Time to move on to other important matters.

Said another way, the answer to why is top management says it is. They make the case for safety.

Is that the only reason why safety is that important? Is that even the real reason why safety is that important?

As you're beginning to suspect, there must be a point to asking those questions. There is. If every leader truly *understood* why safety is that important, every leader would consistently demonstrate that understanding by his or her everyday actions. If every leader in the world over demonstrates that, the world of business and industry would be a better and safer place to work.

Alive and Well at the End of the Day: The Supervisor's Guide to Managing Safety in Operations, Second Edition. Paul D. Balmert.
© 2023 John Wiley & Sons, Inc. Published 2023 by John Wiley & Sons, Inc.

For example, you would be hard-pressed to find anything in the world of work that would look like this:

- Operating equipment isn't maintained as it should be—or as the standards require.
- Known safety problems are ignored because fixing them will take too much time and/or money.
- Unsafe shortcuts taken in order to save money and get the work done fast.
- Poorly trained people operate key equipment where the cost of failure can be catastrophic.
- To satisfy customers, managers don't follow their own safety policies and requirements.
- Safety procedures are reengineered to improve productivity by eliminating checks and balances.
- Significant issues are bottled up by organization silos and not escalated.
- More work is piled on leaders, leaving them with less time to lead and manage safety performance.

These are not made-up examples. Investigation reports found in the public record acknowledged that these exact problems existed in headline making events you may have heard about: Three Mile Island, Union Carbide Bhopal, NASA's Challenger and Columbia, BP Texas City, Deepwater Horizon, Boeing 737 Max, and Rio Tinto Pilbara.

That's simply an observation, not an indictment.

Every one of those organizations placed a high value on safety; their leaders professed a belief in the value of safety. The problem was that actions by leaders in the middle management of those organizations—individually and collectively—belied those values and commitment. It wasn't simply a case of misguided actions by a few leaders. In most cases there were others working in those organizations who knew things weren't right, that operations weren't as safe as they appeared to be, and that actions weren't being taken to support the commitment those words implied. But their knowledge wasn't sufficient to change the course of events.

It's easy to say or write that safety is a corporate value. That doesn't make it a value, one practiced by executives, managers and supervisors as they go about making the important decisions that spell the difference between going home safe or hurt at the end of each day. For example, at Boeing, managers "...chose a path of profit over candor" in withholding critical information from the Federal Aviation Administration about changes to its 737 MAX aircraft that were later found to be factors in two commercial aviation crashes. On Transocean's drilling rig, the Deepwater Horizon, a drilling manager under pressure from a client reluctantly agreed to withdraw drilling fluid from a well that had not been proven to be under control, hoping that if all else failed, the blowout preventer would prevent catastrophe. It did not, resulting in the loss of life, loss of the rig, and an oil spill worse than the Exxon Valdez.

If the leaders in all of these cases had *understood* the real reasons *why* safety is the most important business objective they had, it's likely that things would have been done differently and we might never have heard of any of those names or associated them with tragedy.

TAKE TWO: FROM THE TOP

"Safety will always come first. Period." That's how the top operating executive at Exxon put it, back in the early 1990s.

As you're reading this book, odds are good that you're not the top executive running your company. While executives are one potential audience for *Alive and Well*, this book is primarily written for an even more important audience of leaders: middle managers and front-line leaders working well down the chain of command, often far away from world headquarters. Intel's Andy Grove described these leaders as the "muscle and sinew" of the organization: leaders who, when it comes to managing safety, are the real difference-makers in the outfit.

If you're one of those leaders working in middle management, every day your performance has a far greater impact than that of those in the executive suite on the really important measures of success in your business—how much work gets done, how well the product is made, how well the customer is taken care of, and who goes home safe at the end of the day.

Every leader working in the middle management knows the top executives are the ones who determine strategy, goals, objectives, and set the priorities. The role of middle managers is to understand what those goals are, and then turn around and make things happen. The former is today known as "alignment"—getting aligned with the goals of leaders—and the latter is "execution"—translating goals into results. Those two processes define both the role and the everyday work life of middle managers the world over.

The proper understanding of The Case for Safety—the *real* reasons why safety comes before all the other business objectives—starts at the top of the enterprise with those who set the goals and determine the priorities.

There must be reasons why a CEO running any kind of industrial business puts safety on the short list of goals critical to the organization. What are they? For example, why do you think the top operating executive at Exxon would say "Safety will always come first. Period."?

In their case, there was a simple two-word answer that explained all: Exxon Valdez. That was the name of their oil supertanker that caused what was at the time the worst oil spill in US history. In the aftermath, the company was mired in the predictable debacle of simultaneously trying to clean up the spill and the company's good name and reputation. The call to action by their executive needed no explanation—everyone understood why he said that—and proved instrumental in getting the company on the path to safety excellence.

Fortunately, events like that are rare. If you were to ask a typical CEO, "As the person running the company, why is safety important to you?" the answer would probably go something like this: "Our company is built on delivering great products to our customers, which makes the people who build those products the most important asset we have. We would never want to see any of them suffer any harm. That's why safety is one of our core values."

A good answer, readily understandable. But is that answer "the truth, the whole truth, and nothing but the truth" about why safety must come first?

It's become popular for leaders to profess safety as a personal value, and to identify safety as a value the organization shares. For the moment, we'll assume saying and writing that actually makes safety a value. If only life—and leading—were that simple!

But let's stick with this "values" thing. The theory behind the practice of describing safety as a value is that "values never change". Values last a lifetime and aren't subject to popular trends or the pressure of the moment. The rationale sounds good until you apply just a bit of critical thinking. Then the flaws in the logic become rather apparent.

For openers, where is it written that values never change? Have your values changed over your lifetime? Of course they have. Yours at age 19 were probably a lot different than they are now. Wait till you turn 65. Values do change, and change is often for the better. Over time, without even saying so, the value leaders in business and industry have for safety has changed dramatically for the good. There was a time when human life wasn't all that important to those running businesses.

Perhaps we should take a further step back and define a value; what exactly is a value?

A value is defined as something with intrinsic worth, so important that you need to go no further to know its worth. A value is an intangible. You can't see a value; you place value on something or someone. By that definition, safety serves as the perfect example of a value.

In the twenty-first century, determining and communicating values has become a common business leadership practice. But consider how the practice works—as seen by followers. Followers can't see a leader's values—they are intangible—but they can certainly observe a leader's actions and reach conclusions as to what the leader truly values. That followers routinely do.

Unfortunately, the world is filled with examples where leaders' actions failed to match their professed values.

No matter what value for safety they might claim to have, the everyday actions of leaders powerfully communicate what is seen as their true belief as to the importance of safety. Actions are what followers pay the most attention to, cause them to decide what the value for safety really is, and, in turn, shape how followers decide what really comes first as they do their jobs.

Circle back to the list of conditions and practices that existed within those organizations at the point when the headline making events took place, it's not hard to imagine what followers in middle management directly responsible for the problems thought about *their* leaders' values. Safety did not always come first. Period.

Was that what the top executives really wanted? Unlikely.

So back to where we began. It's possible a CEO's candid answer to the question of why safety is important might go more like this:

> "Look, poor safety performance can wreck this company. A serious accident could cripple production, cost millions of dollars, do irreparable harm to the name and reputation of the company, and cost us customers."

True? Of course it is.

Good safety is good business; conversely, bad safety is terrible business. There's not a thing wrong about being honest about that. In the US alone, the annual cost of medical treatment of work-related injuries is estimated to be in excess of $160 billion, and the loss in productivity is estimated at four times that amount. Suffer a high-profile event, the market value of publicly traded companies can take a nose-dive. As to the upside, stock prices can also reflect the value of excellent safety performance: studies of the oil and gas exploration and production industry and the hydrocarbon pipeline industry demonstrated a clear correlation between environmental safety and stock performance.

In the long run safety is good business, and good for all the stakeholders. Every leader knows that. The problem is with the short run: in the "right here and right now" moment, the safest way is rarely the fastest way, the cheapest way, or the easiest way to run the business—or do the job. Every leader feels that. Over a period of years, doing the right things—investing in training, properly maintaining equipment, buying better tools, planning the work, fixing safety problems—will pay a very good return. The problem is the payout comes in the future, not now. And that investment in safety costs something—time, effort, resources, money—in the short term. An operation can under-train, skimp on maintenance, get by with inferior tools and equipment, and ignore problems and not see any immediate damage from those choices. Worse, leaders are often rewarded in the short term for "doing more with less" leaving their successors to deal with the cost—and the problems.

Yes, there is a business case for safety. Recognizing that requires the investment of "patient money," and maintaining a long-term perspective. Still, that business case is far from the whole truth about why a CEO would take safety seriously. The CEO goes on: "But even if safety weren't good business, sending people home safe is the right thing to do. Those people have lives and families, and I have a moral obligation to provide them with a safe place to work."

That's the ethical case for safety: safety is the right thing to do. Period.

There are CEO's who not only said that, but backed up their words with action. Warren Anderson was one.

In 1984, Anderson was the CEO of Union Carbide at the time of the Bhopal tragedy and his reaction was to quickly travel to the site to survey the damage and see what he could do to bring some relief in the face of the terrible tragedy. Surely the lawyers would have told him, "Warren, don't go. It's too risky."

Despite the risk, Warren Anderson traveled to Bhopal, a small town in India where his company was responsible for the deaths of more than 2,500 people who were living outside his company's chemical plant. Why do that? The only plausible explanation is that his company did something terribly wrong and he felt it to be his moral obligation to accept responsibility for that failure. He must have thought, "What happened was wrong and as the leader at the top, I have to do the right thing to begin to make amends." Books have been written criticizing management's role in the causes of the Bhopal tragedy. No matter: what happened in the aftermath demonstrates one leader's belief in the ethical case for safety.

Warren Anderson's reward for going to the scene and accepting responsibility for the accident was to be placed under house arrest, charged with the murders of those who perished in the accident. He managed to escape the country and escape prosecution, but for the rest of his life, he admitted to being "haunted by what happened in Bhopal." Decades later the Non-Executive Chairman of Union Carbide India Limited at the time of the tragedy, Keshub Mahindra (founder of the Mahindra Company) was convicted and sentenced to two years in prison for "culpable homicide not amounting to negligence."

That leads to a third reason why a CEO might take safety seriously: they have a lot of skin in the game.

Among the many goals that a CEO is responsible for achieving, safety is often on the short list of those most critical for success. Safety performance can determine a CEO's relationship with the board and shareholders, affect status in the business community, compensation, and even tenure in office. The day after his company suffered its thirteenth fatality in less than eight years, the CEO at the mining company Suncor agreed to step down. His company had been under intense pressure to improve safety performance by an activist shareholder.

On reflection, from the vantage point of the CEO the reasons why safety is of the utmost importance are easy to understand: safety is good business; it's the right thing to do; doing it well will reflect favorably on the CEO. There's not a thing wrong with any of those motivations, and any one of the three will cause a leader to do the right thing. They *know* those reasons.

But they still may not *understand* that their reasons aren't the real reasons why safety always comes first. Period.

TAKE TWO: FROM THE MIDDLE

If you manage in the middle of the organization, understanding the reasons a CEO should see safety as a top priority is instructive. It might even prove useful, to the extent that you believe "What's important to my boss is what is important to me." But that's not the only reason—or even the biggest reason—why safety is the most important business objective you've got, no matter what else you're responsible for.

Three simple questions will help you understand why that is the case:

What are the most important things in your life?

If you suffered a serious injury at work, what would be the impact on your answers to that first question?

If you are a supervisor, responsible for the safety of others, do you have any follower whose answers to those first two questions would be fundamentally different from yours?

The answers to those three questions are so obvious, aren't they?

What are the most important things in your life? If you're like the rest of us who work for a living, your job is an important part of your life. Studies show that most people actually like their jobs, and meaningful work is a big part of many people's lives. But not the biggest part. The reasons why we get up in the morning—or afternoon, or evening—and head off to work have to do with things that are more important than what we do for a living. We earn a living to enable us to do the even more important things in our lives: take care of our families, spend time with the people we love, make the world a better place, and yes, have fun.

Those are values: your values. Everyone has values: things that are of the utmost importance. Most values aren't about things; they're about people and living life. Your values are the reasons why you work. It's not the other way around.

If you suffered a serious injury at work, what would be the impact on your answers to that first question? In a word, devastating.

Every day, people die making a living. They're people just like you who, just like you, got up and headed off to work, but never came home. People get seriously hurt. For some in that second group—who are injured seriously but not fatally—the impact is lifelong. Maybe they fell, suffered severe spinal cord injuries, and will live out their lives in a wheelchair or using a walker. If you've worked in industry, it's a good bet you know someone who had something like that happen to them.

If it happens to you, it's not hard to imagine how this affects all the important things in your life: family, friends, finances, and all the things they did for fun. Suffer a serious injury, in an instant life changes. Dramatically. Unalterably. Not for the good. There's no way to turn back the clock and change what happened. Who do you think has the worst of it—you or your family? It might well be the family, the ones who are left to pick up the pieces, live with the impact, and take care of you. Back on the job, the memory of the tragedy slowly fades over time. There might be a plaque placed at the scene, a moment of silence a year later. It's not that way at home, where every day there is a constant reminder: caring for you, paying the bills, and living with all the complexities of having a disabled person under their roof.

Now you've thought about how a serious injury would affect you and all the reasons why you go to work each day. Not that you didn't already know. But that's you. What about the people you're supervising? What would be the impact of a serious injury on their reasons for working? Are they any different from you?

If you are a supervisor, responsible for the safety of others, is there any follower whose answers to those first two questions would be fundamentally different from yours? You know the answer.

Every one of your followers works for exactly the same reason as you: to earn the income to support their values. A serious injury would have the same devastating impact on any member of your team that it would have on you. On that point, they're no different from you.

Not just knowing but understanding the answers to these three questions—the whole truth— will once and for all time change the way you think about your role as the leader responsible for the safety of your followers.

That's The Case for Safety.

YOU'RE RESPONSIBLE

If you're a supervisor, defined as "one who is responsible for the work of others," there's one final, critical aspect for you to consider as part of understanding The Case for Safety: your *responsibility* for what happened.

Picture the case where a serious injury happens not to you, but to one of your followers. As the supervisor you've got certain responsibilities in the situation. You might start addressing them by visiting the job site, where the event happened, to survey the damage firsthand. Not a pretty picture. You may have to call the family and tell them to meet you at the emergency room. Tough duty.

Later, you meet the family face-to-face. You'll hear about the important things in your follower's life: family, friends, interests, and passions. You realize this is really all about your follower's life—and their job is just one small part of that life.

Unlike a top executive, you very well may know the follower and his or her family. They live in your town; you may know the family outside of work. Your follower's family might even be your family.

At some point, you might be questioned by one of their family members, "How did this happen?" You hope you have a good answer, and hope that answer doesn't involve your shortcomings as a leader.

At a public hearing following a mining accident that claimed the life of 12 miners, the mine manager had to face more than a hundred family members. One of them told the mine manager, "When you go home to your family today, think about us not going home to our families. I get to go home to my dad's picture." You can be sure that is exactly what that manager did; it's what you would do.

As bad as that seems, if you're the leader, the worst hasn't happened—yet. On the drive home from the hospital, the next morning, or days later, you'll start asking yourself, "How could I let this happen?" and "What should I have done to prevent this?" You'll begin to examine your responsibility for the damage that was done to someone else's life. That self-examination can be a brutally tough test. There will always be honest answers to the questions you ask yourself; you can only hope they're good.

But what if they're not? That may be the painful truth. If it is, you'll know it. What if you saw your follower not wearing fall protection before falling; you knew about problems with equipment; you knew shortcuts were being taken. What if it was your decision to take the risk, hoping things would turn out well.

Your pain is mental: it's called survivor guilt, pain that you have to live with for the rest of your life. If your follower gets seriously hurt, you can wind up being a victim too. What hurts someone working for you winds up hurting *you*.

That's one lesson you never want to learn the hard way. Best to learn from someone else's misfortune: sadly, there are plenty of examples.

THE CASE FOR SAFETY

Now you *understand* "The Case for Safety." The simple and unchangeable truths about safety: the real reasons why safety matters more than any other business goal, and why managing safety performance is the most important job duty you have as a leader.

- Everyone in your business works to live. It's not the other way around. That's true for you, and equally true for everyone who works for you.

- A serious injury can have a devastating effect on all the reasons people go to work in the first place.

- No matter how important any other business goal might be, it can never justify someone risking the most important things in his or her life.

The Case for Safety isn't made by the top executives when they decide safety is a core value or that it's most important thing. It's not their decision to make. The Case for Safety is made at home; when people decide it's time to head off to work.

Understanding The Case is a sobering thought for any leader. When something goes wrong, it isn't just the follower's life that can be turned upside down; you may wind up being a victim yourself. On the other hand, somebody has to lead, and you've been given that opportunity. No need to spend any time worrying about what really does come first: it's always safety. Period.

Putting safety first isn't necessarily easy. Stand up, speak out when something is wrong, insist that safety problems be dealt with, you may find yourself very unpopular with your bosses and customers, who see the short-term cost of dealing with a safety problem—not the long-term benefit. It takes perspective to see the short-term cost *and* the long-term benefit. The absence of that perspective explains the list of problems described at the start of this chapter. The long-term costs of those cases were colossal; in retrospect, the short term costs were insignificant.

Don't make the mistake of thinking that standing up for The Case for Safety will endear you to your subordinates, either. Sure, some will cheer you on for doing the right thing. Others will complain about your insistence that they work safely, wishing you would just leave them alone to do the job the way they want to do it. Rarely is a supervisor who manages safety performance well and demands compliance given a parade in their honor at the end of the workweek. Leading isn't about winning a popularity contest. It's about doing the right thing no matter what anyone else thinks.

In those difficult times, "The Case for Safety" will serve as your compass: The Case always points to what matters most. Follow it, and it will lead you in the right direction.

CHAPTER 3

THE PRACTICE OF LEADERSHIP

Leadership is the art of getting someone else to do what you want done, because he wants to do it.

—*Dwight Eisenhower*

No follower on the planet wants to get hurt, but without the strong influence and management of safety performance by their leader that simply won't happen. Leaders must lead. The question is how: as the leader, what do you actually *do* to lead your followers to work safely?

The answer to that question is found in the practice of safety leadership. As to finding out what that practice should be, the perfect place to begin is to examine your own practices: what exactly do you do to lead your followers to work safely? If you're a supervisor, manager or executive, no doubt there is a long list of practices you engage in. What exactly are they?

Pose that question to leaders, there's always a long pause, followed by a surprisingly tentative reply. For most, it isn't easy for leaders to explain exactly what they do to lead their followers to work safely.

Rest assured, it's not because they aren't doing much in the way of leading. A more plausible reason is they've been at leading for so long, they largely lead out of habit; habit being behavior performed without conscious thought. Another reason could be that so much of safety leadership seems to be "just part of doing my job."

Complicating matters is confusion about what exactly constitutes "leadership", as differentiated from "management." Not that the lack of clarity stops some from going on at length that what we need are "leaders—not managers." Education as to the difference would certainly help to clear things up. But for those responsible for the safety of their followers, causing followers to work safely is not an academic subject. Practices to lead and manage take place in the real world of the supervisor and manager and executive—and their followers.

The practical solution is to learn by simply watching a good supervisor in action, taking careful note of what the leader does to lead people to work safely.

Alive and Well at the End of the Day: The Supervisor's Guide to Managing Safety in Operations, Second Edition. Paul D. Balmert.
© 2023 John Wiley & Sons, Inc. Published 2023 by John Wiley & Sons, Inc.

A DAY IN THE LIFE

Back in the days when work process reengineering was the rage, this process of direct work observation had a name: "A Day in the Life." If you were interested in improving the productivity of a pipefitter, you followed him around for a day with a clipboard in hand to see what was getting in the way of doing the job. The information learned was used to improve work processes. You'd be amazed at what can be learned just by observation—and paying attention to the details. Watching a leader in action isn't any different: just be sure to pay close and careful attention to everything the leader does—and know what to look for.

Putting the process in play is simple. For example, pick out a good front-line supervisor and follow him or her around for a day. Start when the door to the office is unlocked at the start of the shift, and don't stop watching until the door is closed and the leader heads for the house at the end of a long day.

Do not spare the details: managing safety is all about the details. Every time the leader does something that involves managing safety, write it down. If you aren't sure, write it down anyway, and decide later if it stays on the list.

As it relates to safety, here's how the morning of "A Day in the Life" of this particular front-line leader goes:

6:40 a.m: The supervisor arrives at the office.

6:41: Gets coffee from office pot in the break room, cleaning up a spill left behind by someone else.

6:45: Logs in on the computer and checks e-mail. Notes a report of a near-miss from a crew member, requiring investigation.

6:48: Forwards near-miss report to the Safety Department.

6:51: Writes e-mail to scheduler requesting status of repairs to damaged handrail.

6:58: Receives phone call from the manager, who relates information about a breakdown of critical production equipment that just happened. Repair will be the critical job of the day. Supervisor promises to personally check on the job right after the morning safety meeting.

7:02: Rearranges crew assignments to staff the repair job with highly experienced crew members.

7:16: Chooses the topic for the morning toolbox safety meeting: driving safety; prepares for safety meeting.

7:27: Joins crew in the break room. Asks one about the health of a sick child.

7:30: Starts toolbox safety meeting—on time.

7:39: Asks: "What help do you need to get this job done safely?"

7:41: Reminds: "There is no job we'll do today that is so important that it's worth you getting hurt." (Note: Everyone laughs—the crew hears that every day.)

8:11: Leaves office to visit site of critical equipment outage. Takes hard hat and safety glasses. Hands you, the observer, the same PPE.

8:12: Walks around truck before backing up. Gets in, adjusts mirrors, buckles seat belt.

8:12: Reminds you to buckle your seat belt.

8:13: Drives to job site, following the speed limit. Comes to a complete stop at all stop signs.

8:15: Mobile phone rings: supervisor doesn't answer the call.

8:19: Parks truck; puts on hard hat and safety glasses before getting out.

8:21: Signs in at control room before visiting job site.

8:27: Pauses to watch four crew members at work.

8:28: Observes one member not working safely; decides to intervene.

8:29: Compliments three for full compliance with the PPE requirements.

8:30: Asks fourth member of crew: "Why aren't you wearing your hearing protection?"

8:31: Listens to the explanation: "I just forgot."

8:32: Explains the hazards present at the site, the risk of injury, and the consequences of not following the rules.

8:35: Asks crew questions about the progress in getting scaffold erected.

8:36: Listens to a problem with getting the scaffold inspected before use.

8:38: Advises crew: "Let me see what I can do to help. I'll get back to you."

8:39: Radios scaffold inspector. Secures commitment for prompt inspection.

8:45: Communicates that information back to the crew.

It seems like an ordinary start to the day in the life of a leader; in many respects it is. In a little over two hours—beginning well before the official start of the shift at 7:30 a.m.—we've filled a page full of activities, providing a huge amount of useful information about the practice of safety leadership, not in a theoretical sense, but what it looks like in the real world and real time of operations.

The list is chock-full of simple stuff. So simple that most of the items might not even strike you as leadership in action: asking questions, checking on work, showing a genuine concern for employees, fixing problems, reminding people to follow the rules, following the rules yourself, closing the loop by giving feedback about the status of work. As to the location of these activities, they take place out on the job, back in the office, in the break room.

Sound familiar?

Sure it does: you find yourself doing things like this every day. They hardly seem like the great moments of leadership, those special times when the hearts and minds of followers are seemingly won over to safety: no memorable speeches; no monumental decisions; no great event.

That's an important point to understand about safety leadership: the real stuff of leadership is found in the ordinary, everyday activities of supervision. It's not that the big things don't matter, but they just don't happen very often. Whereas, those small, easy to overlook things happen all the time, and they do matter—to followers.

Most of the time, good leadership looks downright boring. That powerful truth about leadership is often missed by the experts. Pick through the popular books, models, and theories, and you'll find that many authors, teachers, and consultants succeed only in complicating the process of leadership. In practice, safety leadership is pretty simple.

Just don't confuse simple with easy.

Understood properly, this list of the leadership activities undertaken by one supervisor is both basic and impactful: explaining, solving problems, giving feedback, checking on things. Bear in mind, this list is just the tip of the iceberg: were you to watch a good leader in action over the course of a full day, or a week or a month, you could easily come up with a hundred things he or she does to lead people to work safely, from equipment inspections to performance appraisals: scheduling training, passing out safety awards, counseling an employee who can't seem to follow the safety rules, and so on. Those seemingly mundane tasks might not make for a best-selling book on leadership—but they are exactly what make the difference in sending people home safe.

We've looked at a front-line supervisor in operations; we just as easily could have chosen a manager or an executive. The specific activities on the list would look a lot different: rarely does a top executive run a toolbox safety meeting at the start of the shift, but they all hold regular staff meetings. Still the underlying principles would be the same.

You would do well to undertake this exercise for your practice of safety leadership: you might be surprised as to what you'd learn from the process.

LEADING VERSUS MANAGING

In management science, there is indeed a difference between leading and managing: supervisors, managers and executives are required to perform both functions. Let's use our list of this leader's activities to clear up that misconception about leadership, one that's often expressed as "We need leaders—not managers!"

We'll begin with questions: How would you characterize the relative importance of giving a follower a compliment versus following up on the status of a handrail repair? Giving advice about priorities versus assigning the right crew members to a critical job? Stopping an unsafe job or giving a motivational speech at a safety meeting? Which of these activities do you think contribute more to seeing to it that everyone works safely?

The simple answer would be they all matter.

For simplicity's sake, we'll call this list "Leadership Activities." Some might be considered leading and others managing; all are integral to sending people home alive and well at the end of the day. As a practical matter, they're inseparable, all part of the work of the leader; of being a good leader. In that sense, the difference between leadership and management is a distinction without a difference.

That said, in academic terms, there is a difference between leading and managing. The lack of understanding of that difference has contributed to the mistaken belief that managers aren't leaders. Let's clear up that misunderstanding, explaining the work of the leaders, from front-line supervisor to top executive.

Simply stated, there are four functions of management: planning, organizing, leading, and checking/evaluating/correcting. If that sounds a bit like Deming's Virtuous Cycle, "plan/do/check/correct," it would not be a coincidence. That cycle of continuous improvement neatly describes what supervisors and managers do, and it's a very useful way to classify their work.

Examples of all four of these management functions being practiced—and performed well—by our front-line supervisor can be found in our list of activities taking place in the first couple of hours in a typical day:

- Planning: preparation for the morning's safety meeting.
- Leading: praising, correcting behavior, giving advice, asking about a follower's ill child.
- Organizing: Assigning senior, experienced crew members to a critical job.
- Checking/evaluating/correcting: Following up on a repair, reviewing a near-miss report, checking on the status of a planned inspection, giving follower's feedback about the status of work.

The chapters of *Alive and Well* can also be organized in this way. For example, Chapter 17, "Measuring Safety Performance" treats one element of the function, checking/evaluating/correcting. Understanding What Went Wrong treats another, evaluating and correcting problems. Chapter 16, "Investing in Training" is an example of the planning function applied to managing safety performance. The majority of the chapters, including this one, "The Practice of Leadership", focus on various aspects of the leadership function.

Every supervisor plans, leads, organizes and corrects. That now properly understood, the value of leadership cannot be overemphasized: leadership is of the utmost importance. But sending every follower home, alive and well at the end of the day requires planning, organizing, and a lot of checking.

Leaders must lead—and manage.

WORDS AND ACTIONS

Now that we've cleared up the academic science of the four functions of management, let's turn our attention back to practical matters, and what well may be the most important thing to be learned about the practice of safety leadership. It applies to leadership wherever it is practiced. The list of leadership activities provides the perfect means to do so.

We constructed our list chronologically, writing things down in the order in which they occurred. That's how leadership happens in real life: events drive leaders, providing the occasion to lead. As we've already shown, these activities can be sorted into the four functions of management. But an even better way to sort them is into two categories, words and actions, as shown in Figure 3.1.

LEADERSHIP ACTIVITIES

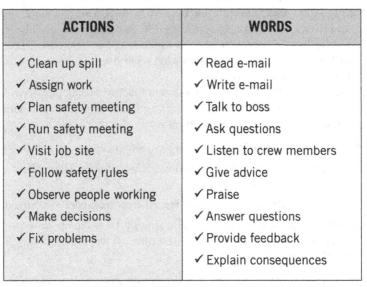

ACTIONS	WORDS
✓ Clean up spill	✓ Read e-mail
✓ Assign work	✓ Write e-mail
✓ Plan safety meeting	✓ Talk to boss
✓ Run safety meeting	✓ Ask questions
✓ Visit job site	✓ Listen to crew members
✓ Follow safety rules	✓ Give advice
✓ Observe people working	✓ Praise
✓ Make decisions	✓ Answer questions
✓ Fix problems	✓ Provide feedback
	✓ Explain consequences

Figure 3.1   Leaders lead by "using words" and "taking action".

This categorization and the names given to the two columns provides a very powerful insight into the practice of leadership. It's one that leaders and followers alike intuitively know, but don't normally identify as the practice of safety leadership.

The distinction between the left and right columns is simple, and hardly arbitrary. The activities shown in the left column involve *action*—what the leader physically does: cleans up a spill, assigns work, makes a decision, solves a problem, wears PPE, follows the safety rules.

By comparison, everything in the right column involves some form of *communication*. On the one hand, the process of communication between us humans is an enormously complex one, often fraught with misunderstanding. Reflect on that process and the underlying human technology of language and you can begin to see why that is so.

Language is tough stuff, but there's one thing common to most forms of communication: *words*: Words can be spoken or written. In the practice of leadership, words can come from the leader to the followers, or from a follower to the leader. In either situation, listening is just as important to the process as is speaking.

Despite all the complexity that has been brought to the process of leadership over the years, leadership really is that simple. Leadership by any leader at any level and in any walk of life always boils down to "using words" and "taking action." Moreover, it doesn't seem to matter who the followers are, or the kind of work or activity they are engaged in, or the level in the hierarchy of the leader.

The elegant simplicity of the practice of leadership takes us back to our Leadership Model, where we defined a leader as someone who has followers. The model can be expanded by pointing the arrow up *and* down: followers create leaders by following, and leaders complete the cycle by leading through their words and actions (see Figure 3.2).

The Leadership Model

Figure 3.2   The Leadership Model provides a simple and powerful way to understand the practice of leadership.

This simple model speaks volumes about the process and practice of leadership. Leadership really is all about the followers—not the leader—who ultimately determine the success of their leader. Understanding that to be the case, the test of the effectiveness of your leadership practices must be, "What works best to influence my followers?"

To determine the answer, you must always see things—and yourself—from the perspective of your followers.

LEADING BY EXAMPLE

Viewing leadership from the perspective of the follower begs an important question: of the two leadership activities—words and actions—which has the greater impact on followers? Influence is the point of leadership, is it not?

The answer is obvious: actions. Every follower knows "actions speak louder than words."

When it's the follower's turn to lead—every leader is a follower of some other leader—they sometimes forget that. In the heat of battle the words can seem so important. That causes leaders to focus on making speeches, writing letters, issuing new policies, rallying followers to the cause. The followers always know better: talk is cheap. Actions matter more.

No doubt you intuitively know action speaks louder than words, but not everyone understands why that is so.

The power of action stems from the fact that we humans are first and foremost a visual species. Long before we had language, we survived by sight—and our other senses. Of the five senses, our power of sight is our most powerful. Thanks to the technology of brain imaging, we now know how powerful our sense of sight is: more than 30% of the active memory capacity of the human brain is devoted to processing visual stimuli. So, action speaks louder than words because action plays to the visual side of our brains: action is what people can see. As listeners—and followers of leaders—we're wired for sight.

Understanding that suggests the most powerful thing a leader can do to influence followers is to lead by example. No matter what the leader might *say* about safety, the visual *image* of working safely, following all the rules and fixing problems sends a far stronger message.

From our list of leadership activities, consider ones such as these: checking the truck by walking around; adjusting the mirrors, buckling the seat belt, stopping at the stop sign, putting on PPE before getting out. They are **Leading by Example**: it is as simple as that. All are 100% visual to any follower who might find themselves in a position to observe the leader's actions.

All the safety rules are there for a reason: to keep people safe, including the leader. But seeing the leader following the safety rules also sets a powerful visual example for followers. No one should be better at following the safety rules than the leader.

That may be simple common sense, but it is hardly common practice. Many leaders act as though the rules are for everyone else's benefit, not theirs. And they seem to operate under the presumption that no one will notice if they don't follow the rules. But that's often when followers pay the most attention.

Leading by bad example is still leading; just not leading followers in the direction the leader wants them to head.

It's always tempting—and usually easy—to find fault with other leaders. What matters is *your* practice of leading by example. Sure, just about every front-line supervisor will show up at the job site wearing a hard hat and safety glasses. But what about the drive out to the site? Did you buckle your seat belt? Come to a complete stop at all the stop signs? Drive at the speed limit? Ignore the incoming calls on the mobile phone?

These are seemingly small things. If you're a leader, that is yet another important lesson to be learned about the practice of leadership: followers pay far closer attention to the small examples of their leaders than most leaders believe they do. That fact can be very annoying to the leader who's thinking, "I'm not perfect. That's impossible, and I should be allowed to be me."

Yes, being perfect is impossible. But once a follower identifies someone as their leader—remember the Leadership Model—they look at that person in a different way: "You are my leader." That is what followers do to their leaders. It is as if they place the leader on a pedestal, to be looked up to.

Followers expect their leaders to be better role models than themselves. The best leaders are.

USING WORDS

Now you understand why a leader's actions count more than words: action is visible to followers, who are by their nature visual creatures. That should come as good news for every leader: it means those seemingly small actions such as putting on the PPE and cleaning up a spill in a break room play large. That actually makes safety leadership a lot simpler.

Actions do speak more powerfully than words; still using words is significant and important part of practice of leadership for words. To take full advantage of the power of words, it helps to understand exactly what a leader uses words for. On that point, "to communicate" is not a very useful answer. Words perform four vital leadership functions: explain, elicit, excite, and engage. These four functions cover a wide swath of key interests of a leader:

Explain: Words are the means of performing many basic supervisory duties like assigning work, training, giving feedback, and communicating change. Words can explain the actions of a leader.

Elicit: Words are a way to elicit feedback from followers. Eliciting gets the follower talking. That creates a dialogue, puts a leader in a listening mode, and enables the leader to learn what followers are thinking.

Excite: Words have the power to excite followers. Excitement can be a very positive force: motivated followers can move mountains. But the capacity of words to excite cuts both ways: a leader's words are also equally capable of de-motivating followers.

Engage: Words can serve as a means of engagement, the antidote to complacency. Complacency means little thought is being given to the task at hand. Engagement is entirely different: followers are focused, paying attention, and actively thinking about the matter of safety.

As a leader, having the right words to perform each of these four functions proves very useful. One example is the use of the Stump Speech. A Stump Speech is a way to organize your words as the leader, and to be prepared to put them in play when the occasion presents itself.

Let's begin with a definition of this leadership practice: a Stump Speech is a concise statement of the supervisor's personal beliefs, values, advice, and expectations about working safely. If you've been around any organization for very long, chances are you've been on the receiving end of what you would recognize as a Stump Speech from a leader you've worked for.

Here's an example of a good safety Stump Speech, one regularly employed by a Plant Manager in a large chemical plant.

Nothing we do can ever become more important than our safety performance.

At the end of the day, if we don't make our product, if we miss a shipment, if we run over budget, if we send the customer away unhappy, those things only cost us money. For that, we always have tomorrow to try to make up what we lost today.

Safety's not like that. We can't replace human life, and we never get a second chance to do it right.

That's why safety always comes first.

Having a Stump Speech means you've taken the time to think about what you have to say on an important aspect of working safely. You don't want just one Stump Speech, but rather series of speeches on subjects for which you have a take. Here are several examples:

- Why safety always comes first?
- Why to stop a job that isn't safe?

- Why to report near-misses?
- Why to follow all the rules all the time?
- What to do if you can't?
- When to be extra careful?
- Why to say something to a peer who isn't working safely?

With a Stump Speech in hand, when the appropriate time arises for you to offer your good advice and counsel, you'll know the message you want to deliver. A Stump Speech is a far better alternative than trying to come up with something brilliant in the moment. You just need to be sure it doesn't sound like something that you memorized and are repeating by rote.

As to timing, your Stump Speech can be used on any occasion when you're naturally looked upon as the leader, such as:

- Safety meetings
- Welcoming new members to your team
- Interviewing prospective employees
- Introducing new safety policies and procedures
- Dealing with problems and the "crisis" of the day
- Kicking off an accident investigation

Giving a Stump Speech to your followers might strike you as preaching. In a sense, it is. Followers want to know what their leaders think, and they listen to what their leaders have to say. Having the words in the form of the Stump Speech make doing that easier.

Keep in mind the purpose of a Stump Speech is to influence followers. Influence starts with the content of the message: the advice, philosophy, values, and expectations given by the leader. Followers know good content when they hear it. They also are perfectly capable of recognizing "corporate speak" when they hear it. The words the leader uses need to express the content in a way that connects with followers so they understand the message, and see it as having meaning and value—for them.

While calling this message a "speech" might suggest length and formality, the best Stump Speeches are short and to the point. But there's a huge difference between a speech that is concise and one so lacking in useful content that it's little more than a slogan. A speech is not a slogan. By definition, a slogan lacks content; it is the content—beliefs, values, advice, expectations—that makes your message a speech. Your content need not be long; it just needs to be seen as good.

Then there is the matter of what is called delivery: how the words are communicated. You might think good content coupled with effective delivery would be the recipe for speaking in a way that guarantees success in winning followers over. The truth about effective communication is that two other factors have greater influence than content or delivery.

The first is how the leader looks when speaking: body language has been shown to have greater influence in communication than either content or delivery. Given what is known about power of sight for us humans, the finding makes perfect sense.

The second factor is even bigger: credibility. No matter how good the body language, or for that matter the delivery and content, whenever a leader attempts to influence by using words, their credibility will be the difference that makes the difference. That makes it imperative that a leader properly understands something that "everyone knows" about credibility: where a leader's credibility comes from.

From action, of course.

That takes us right back to the action side of leadership activity. Credibility is based on what followers know about the leader's prior record of action. A record of proper action gives a leader credibility and will lead followers to listen and be influenced. When a leader is lacking credibility, followers will tune out the leader.

That means in a very real way followers determine a leader's credibility. That a leader must establish credibility is one more example of how followers play such an important role in the process of leadership. Credibility is something earned by the leader—from followers.

Does the leader's position in the hierarchy impact their credibility—and influence?

The obvious answer is yes, but how so? Intuitively, it would seem that the higher in the organization, the greater the leader's influence on followers. But, as will be explained in subsequent chapters, there is a difference between a leader's formal power—such as the scope and size of their approval authority—and informal power—the capability to move the needle of influence—to change behavior. When it comes to the exercise of influence, surveys consistently show that the front-line supervisor is the most credible member of management as viewed by those doing the work.

That should come as no great surprise to anyone with any experience working in a large organization. An executive might look like someone who wields great influence when speaking to a large gathering, but as a practical matter, the supervisor is the one holding the real power to influence when speaking to a small group of followers. This means there is no one better suited to make a Stump Speech for safety to the crew than their front-line leader.

Finally, there is the question of repetition: does a Stump Speech need to be kept fresh to keep follower's interest?

It might seem that way, but there is great value in giving a consistent message. And studies have shown it takes repeating a message at least six times for the listener to even remember it.

Don't be reluctant to give the same speech over and over. Just be sure every time you do, you look like you mean what you say.

THE PRACTICE OF LEADERSHIP

More than a half century ago, Peter Drucker wrote the first book on the profession of management, *The Practice of Management*. To this day, it's still the best book written on the subject. A brilliant observer of management and leadership, Drucker was of the opinion that, as important as leadership was, it was something that could neither be taught nor learned.

That might strike you as rather odd: here's a brilliant management consultant who made his mark as writer and teacher, who did not think leadership could be taught. The foundation of his logic makes perfect sense. In his practice, Drucker observed all kinds of successful leaders displaying entirely different personalities and proficiencies. So much so that he wasn't able to correlate any of that with success as the leader.

The one thing that Drucker could correlate with success was "integrity." That was the word he used to describe what he saw as the defining characteristic of effective leadership. It was his view that integrity could not be taught or learned; hence his conclusion.

Today there is tendency to think that effective leaders are all stamped from the same mold; the truth is that the best at leadership are an incredibly diverse lot, with a widely divergent set of core competencies. It is a mistake to think otherwise, or to try to be someone you are not.

But to Drucker, *practices* were an entirely different matter. Leadership practices are what leaders actually *do* to lead. As Drucker put it so well: "Practices, though humdrum, can always be practiced no matter what a man's aptitudes, personality or attitude. They require no genius—only application. They are things to *do* rather than talk about."

Over the course of this chapter, having closely examined effective safety leadership in practice, you now appreciate that practices of the type described by Drucker are found in the words and actions of the leader. No practice is more important than Leading by Example. As for words, the Stump Speech represents a very good place to begin your dialog with your followers. That's the starting point for the practice of leadership. You'll find many more of "those things to do rather than talk about" in the succeeding chapters of *Alive and Well*.

Are those leadership practices humdrum? Certainly.

Are those humdrum practices the stuff of great safety leadership? Absolutely!

CHAPTER 4

MOMENTS OF HIGH INFLUENCE

Carpe diem. [Seize the day]

—*Old Latin proverb*

Leading, and leading well, demands the most precious resource a leader has: time. When it comes to sending everyone home safe every day, there is no substitute for the time and attention of a leader. Time is the one resource that a leader can't buy more of: there are only so many hours in the day. Leaders—supervisors and managers who are running the business—are some of the busiest people on the planet, with production, cost, quality, customer, reliability, business process improvement constantly competing for their time. If you're a leader, how do you find the time it takes to lead people to work safely?

You could try the time-management trick of finding things to *stop* doing. Prioritize the important things: do only them and ignore the less important stuff. It sounds like a great idea—until you put it into practice. Try not doing something that's important to your boss, or your customers, or your followers: how long do you think it will take them to notice? If you ask them what you might stop doing, you're more likely to come away from that conversation with even more to do. Everything you do is important to someone else.

There is an alternative, a better answer to the question, "With so much on my plate, how do I find the time to lead safety?" You already have plenty of time to lead—if you consider the time you're *already* spending leading people to work safely. How much time is that? Leaders typically spend far more time leading people than they think they do. They just don't appreciate all the things they do that constitute leading, in large part because those things look like the ordinary, everyday work of the leader.

If a leader doesn't recognize *when* they're actually leading their followers, they'll be missing out on *how* they are leading their followers.

Here a perfect illustration.

Alive and Well at the End of the Day: The Supervisor's Guide to Managing Safety in Operations, Second Edition. Paul D. Balmert.
© 2023 John Wiley & Sons, Inc. Published 2023 by John Wiley & Sons, Inc.

Monday Morning, 6:30 a.m.

An hour before day shift officially starts, a senior supervisor responsible for a crew of twenty-five people is sitting in the office, preparing for the Monday morning toolbox safety meeting. Scrolling through the emails, looking for a potential topic for the safety meeting, it doesn't take long to spot the topic du jour. It's a directive, straight from the head office, with an all too familiar title, "Critical Safety Bulletin." The exclamation point for the e-mail really wasn't necessary.

Seems as though these are a weekly occurrence. This one comes on the heels of a serious incident at another facility. While the investigation is still a work in progress, a significant flaw was uncovered in a key, company-wide safety procedure. Hence the e-mail, and the accompanying directive: a change in procedure "…effective immediately. Please communicate to all affected by the change."

In this supervisor's case, every member of the crew.

The supervisor grumbles, "Another knee-jerk over-reaction. Instead of just dealing with people who can't, or won't …" The leader realizes there's no sense going any further down that path. Still, mood soured, tone set for the week, and facing one more tough safety challenge—selling what will be an unpopular change to a hostile crew—the supervisor decides the best thing to do is to head down the hall for coffee. Perhaps a little caffeine will help.

The office coffee pot is situated in the break room, the location of the safety meeting. At 6:40 a.m. it should be deserted, but not on this Monday morning. Perched in front of the coffee pot happen to be the two most senior members of the crew of 25. If the supervisor had given it a moment's thought, it was all so predictable: these well-known "early birds" always show up, get their coffee, and plunk down in their favorite chairs at the far table.

But on this Monday morning, the leader was preoccupied with how to sell a change in procedure.

"Good Monday morning, Boss." one of them says, with a smile. The leader does enjoy a fine relationship with the crew.

The leader doesn't miss a beat: "'Good Monday morning?' You wouldn't be saying that if you read the news I just read coming from HQ. Another one of those changes coming down from on high. 'Good grief!' would be closer to the truth."

That does not escape notice from the second follower: "Well, I guess that's why you get paid the big bucks." He smiles: gotcha … in a good way.

The supervisor can't help but crack a smile, too.

Coffee cup filled, the leader says, "See you at 7:30" turns and heads back down the hall to the office.

Start to finish, the whole interchange took 37 seconds.

7:30 a.m., Safety Meeting

Fully caffeinated, communication plan devised, game face on, that supervisor is now standing in front of a room filled with 25 followers, including the 2 senior members, sitting in their favorite chairs at the table in the back of the room.

"Good Monday morning, all. I trust everyone had a fine weekend. I most certainly did. Now it's time to get back to the serious business of running our operation safely.

I want to inform you that there has been an important change in a procedure, one that affects the safety of every one of you in this room. Owing to serious incident that happened at one of the company's sites, the new procedure for …"

The leader explains all the details involved in the change. Hands out a copy of the revised policy and finishes in fine style: "Now, I'm confident that each of you not only understands this change, but accepts it, and will help me make it happen.

Are there any questions?"

MOMENTS OF HIGH INFLUENCE

Let's give a name to what happened over the coffee pot almost an hour before the official start of the shift: it was a Moment of High Influence. These Moments are defined in this way: in the everyday life of a leader, there are times and places in which the leader is in a natural position of leadership and followers are in a high state of readiness to be influenced.

As you begin to appreciate the practical application of this leadership tool, you will see the conversation over the coffee pot wasn't the only Moment of High Influence in this scenario. A Monday morning toolbox safety meeting is also a Moment, albeit not nearly as significant as the conversation between the leader and two senior followers over the coffee pot. Communicating a change in policy also creates a Moment, even if it generates resistance to the change. Moments define *when* followers are inclined to sit up and pay attention to what their leader says or does. Communicating an unpopular change guarantees a Moment will happen; getting resistance only indicates the followers are not ready to go where the change says they must. So, don't confuse a moment of great resistance with a Moment of High Influence.

Moments signal to the leader "game on." There is no better time to lead than when followers are engaged; what the leader does in the Moment constitutes leading, when followers are paying attention.

It's all so simple. The way to solve the challenge of time is not to prioritize and eliminate but instead to make the most of your Moments of High Influence. It's the perfect illustration of productivity: don't do more, do better.

But as simple as this tool of the Moment of High Influence is—it really is that simple—it requires the leader to *recognize* the moment. Therein lies the challenge: Moments are not always what the leader thinks they are.

That's because followers determine the Moments. Leaders think it's the other way around: they determine the Moment. That's wrong. Followers are the ones who decide to pay attention, and how much attention they will pay: a little or a lot.

In that sense, Moments of High Influence can be put on a scale: high, higher and highest. In the scenario involving communicating a change in policy, three Moments can be found. There is the Monday morning toolbox safety meeting, but a routine safety meeting isn't nearly the Moment announcing a change would be; the informal conversation over the coffee pot is the biggest Moment of the three. As to what made it so big, on the one hand it was brief and involved two followers, before the start of the shift. But it was a huge moment in that the leader revealed their true feelings about a change in policy. You can be sure by 7:30 a.m. every follower knew full well their leader didn't like the change coming down any better than they did. They just didn't know what the change was.

As obvious as that might be to you, it wasn't for the supervisor. Yes, the leader recognized the safety meeting as a Moment and came in to work early to prepare.

When the change became the topic of the meeting, the leader invested time preparing to communicate the change. Those Moments were recognized. But the leader missed the biggest Moment of all—because it didn't look like a Moment. It was just one of those casual conversations leaders have with followers all the time.

That's how the leader saw it—but not the followers; they determine the Moments.

MISSING THE MOMENT

Missing the Moment of High Influence makes leading tougher and can mislead followers by sending an entirely unintended and wrong message. Here's an example:

Picture an annual site visit by the company president, who delivers a wonderful speech on safety to an audience gathered in the conference room. But on the drive in to the meeting room, the executive complained about valuable time being wasted, being required to sit through the safety orientation video at the entrance point. During the drive, the seat belt wasn't buckled; oh yes, the executive took an important phone call from an investment analyst en route. The hard hat and safety glasses were left in the car.

The executive was certain the speech was the Moment of High Influence: from the front of the room, it was evident every follower in the audience paid rapt attention to the words spoken. Moments later, the speech will have been forgotten; years later, followers will still be talking about what was said in the reception area, and the action of the conspicuous noncompliance on the drive in.

That's the thing about Moments of High Influence: the biggest ones can come when a leader thinks followers aren't paying attention.

There's a saying, "No leader ever has an off-camera moment." That can be a source of frustration to leaders: "Don't I get to act like a normal human being?" The short answer is "No." Remember the Leadership Model: once your followers see you as their leader, you are always their leader. It doesn't matter where you are, or what the topic of conversation.

Here's the good news in that: once you recognize the power of your Moments of High Influence, leading becomes easier. You're going to handle all those everyday situations that are your Moments in some way: why not handle them the best way possible.

You're a lot more likely to do that when you recognize them for what they are: your Moments of High Influence.

MOMENTS HAPPEN ALL THE TIME

What are the Moments of High Influence for your followers, and therefore for you as the leader? Looked at in the rearview mirror, a casual conversation over the coffee pot, turns out to have been a huge Moment. So would be a serious incident, the violation of a life critical safety procedure, or celebrating a significant safety milestone like an injury-free year. Moments like those are obvious to the leader.

In Chapter 3 we followed a front-line supervisor around for the morning, looking for what he or she did to lead his or her followers to work safely. It didn't take long to fill up the page with leadership activities. Now the question is: were there any Moments of High Influence?

6:40 a.m.: The supervisor arrives at the office.

6:41: Gets coffee from office pot in the break room, cleaning up a spill left behind by someone else.

6:45: Logs in on the computer and checks e-mail. Notes a report of a near-miss from a crew member, requiring investigation.

6:48: Forwards near-miss report to the Safety Department.

6:51: Writes e-mail to scheduler requesting status of repairs to damaged handrail.

6:58: Receives phone call from the manager, who relates information about a breakdown of critical production equipment that just took place. Repair will be the critical job of the day. Supervisor promises to personally check on the job right after the morning safety meeting.

7:02: Rearranges crew assignments to staff the repair job with highly experienced crew members.

7:16: Chooses the topic for the morning toolbox safety meeting: driving safety; prepares for safety meeting.

7:27: Joins crew in the break room. Asks one about the health of a sick child.

7:30: Starts toolbox safety meeting—on time.

7:39: Asks: "What help do you need to get this job done safely?"

7:41: Reminds: "There is no job we'll do today that is so important that it's worth you getting hurt." (Note: Everyone laughs—the crew hears that every day.)

8:11: Leaves office to visit site of critical equipment outage. Takes hard hat and safety glasses. Hands you, the observer, the same PPE.

8:12: Walks around truck before backing up. Gets in, adjusts mirrors, buckles seat belt.

8:12: Reminds you to buckle your seat belt.

8:13: Drives to job site, following the speed limit. Comes to a complete stop at all stop signs.

8:15: Mobile phone rings: supervisor doesn't answer the call.

8:19: Parks truck; puts on hard hat and safety glasses before getting out.

8:21: Signs in at control room before visiting job site.

8:27: Pauses to watch four crew members at work.

8:28: Observes one member not working safely; decides to intervene.

8:29: Compliments three for full compliance with the PPE requirements.

8:30: Asks fourth member of crew: "Why aren't you wearing your hearing protection?"

8:31: Listens to the explanation: "I just forgot."

8:32: Explains the hazards present at the site, the risk of injury, and the consequences of not following the rules.

8:35: Asks crew questions about the progress in getting scaffold erected.

8:36: Listens to a problem with getting the scaffold inspected before use.

8:38: Advises crew: "Let me see what I can do to help."

8:39: Radios scaffold inspector. Secures commitment for prompt inspection.

8:45: Communicates that information back to the crew.

In the first two hours of a busy, but normal morning, there have already been five Moments of High Influence:

- The conversation about the health of a follower's family member
- The toolbox safety meeting
- Observing the work practices of four crew members
- Giving feedback—both positive and corrective—to those members
- Hearing about—and addressing—a safety problem

Dig deeper and you might consider several more Moments of High Influence—if you assume that a team member observed the supervisor:

- Cleaning up the coffee spill
- Complying with the safety rules for driving
- Reminding you—the observer—to buckle your seat belt

TEN MOMENTS OF HIGH INFLUENCE

✓ New hire safety orientation

✓ New supervisor, first day

✓ Near miss

✓ Safety Suggestion

✓ Crisis

✓ Communicating a new policy

✓ Celebrating a milestone

✓ Supervisor visiting a job

✓ Bringing a problem to the supervisor

✓ Investigation

Figure 4.1   Moments of High Influence are frequent and often predictable.

Now you're beginning to appreciate that Moments happen all the time.

If you're a leader, the news gets even better: a leader doesn't have to do anything to create many of these Moments. What goes on in the everyday life of an organization creates the Moments for the leader: shifts start, conversations occur, problems crop up, leaders are called upon to follow rules that apply to everyone else. (See Figure 4.1.)

But if you're looking for them—you should—you should know that Moments predictably arise from four sources:

- The leader can create a Moment. The "coffee pot Moment" is an example.

- A follower can create a Moment. Whenever the follower creates the Moment—such as by bringing up a problem—it is always a significant Moment because it comes from the follower.

- Events create Moments—for example, an incident or near-miss. Events can be positive: celebrations and accomplishments.

- Procedures. A routine procedure like a new hire safety orientation is usually a significant Moment as that's when new followers have their first exposure to safety at their new job.

Moments of High Influence makes leading a lot simpler—time to lead!—but don't confuse *simple* with *easy*. Recognizing the Moment is easier said than done. All too often a Moment of High Influence is overlooked in the heat of battle because it looks like a moment of "that happens all the time." So the new hire gets shuffled off to a coworker for the job orientation; the safety meeting winds up as a discussion on the latest business problem; compliance problems get ignored because confronting people is uncomfortable; the coffee spill gets ignored because cleaning it up is someone else's job.

As you move up in the organization chart, the situation gets worse, not better. The higher in the organization the leader is, the more people pay attention to the leader *in the moment*. If senior leaders understood that, there are many instances in which their behavior would be entirely different. For example, realizing its impact, the CEO would be cheerfully sitting through the safety orientation and dutifully following the safety rules at the site, knowing that would have spoken far more powerfully as to the commitment to safety than anything shown on the powerpoints used in the presentation.

That said, it's always easier to find fault with others, particularly when they're our leaders. As a leader, you don't want to make the mistake of missing your Moment of High Influence. Recognizing them becomes easier when you understand that your moments invariably arise from four causes or sources: as the leader, you can create the moment; your follower can create the moment; an event can create the moment and executing certain processes and procedures such as holding a new hire safety orientation can create the moment. Examples of all four of these sources of Moments of High Influence can be found in the examples shown in Figure 4.2.

When you find yourself in the middle of these everyday situations that meet the test of a Moment of High Influence, you want to handle them in a way that advances your cause, sending people home safe at the end of every day.

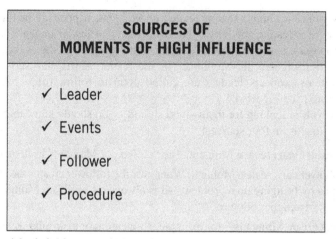

Figure 4.2 A leader's Moments of High Influence come from four readily identifiable sources.

CARPE DIEM

There's an old Latin saying that goes "Carpe diem" ("Seize the day"). It perfectly sums up the leadership tool of the Moment of High Influence: there is no better time to lead in than in your Moments.

Recognizing the Moment is half the battle; the other half is knowing what to do as the leader to take full advantage of the Moment when your followers are paying attention. That's what the rest of this book will tell you: what to do, and how to do it to make the most of your Moments.

Seize the moment—the Moment of High Influence!

CHAPTER 5

MANAGING BY WALKING AROUND

You can observe a lot just by watching.

—Yogi Berra

When it comes to sending everyone home alive and well at the end of every day, there is no substitute for the presence of the leader on the job, where the work of the business takes place. Every leader on the planet knows safety performance can't be managed from the confines of the office, and every front-line supervisor understands one of their most basic duties is the observation of their followers as they perform their jobs.

But knowing that to be so doesn't always mean that's where the leader is most likely to be found. The competing demands on a leader's time—meetings, phone calls, conference calls, e-mail, forms—all too often push spending time out on the job well down the daily to do list. The urgency of those other matters requires a leader to focus on them. By comparison, time spent simply being "out on the shop floor" doesn't seem to produce any tangible results. That makes the practice feel as though it's something that a leader can engage in when there is the luxury of time.

There will never be enough time. That reality serves as the premise on which the incredibly valuable tool called the Moment of High Influence is founded.

Were a leader to fully understand and appreciate the benefits—and the need— for their presence on the job, the investment of the time to do that would always be seen as a high priority.

ON BEING SEEN

To begin to appreciate the benefits to be found in this simple and basic leadership practice, put yourself in your follower's shoes for a moment: when you show up on their job, what do you think is going through their minds?

Alive and Well at the End of the Day: The Supervisor's Guide to Managing Safety in Operations, Second Edition. Paul D. Balmert.
© 2023 John Wiley & Sons, Inc. Published 2023 by John Wiley & Sons, Inc.

"Here comes the boss. What's up with that? Is there a problem?" Not only does this thinking reflect a Moment of High Influence has been created by you as the leader just by showing up, this initial reaction demonstrates the power found in your physical presence. You are always *seen* as their leader.

Any time you see someone putting on some piece of missing PPE because they see you coming, it means there's an even bigger Moment in play. In a perfect world, that wouldn't happen because everyone would work safely without the need for the presence of the leader. But in the real world, that's not how this works.

Moreover, when people change their behavior for the better because they see you coming, safety performance never gets worse.

Looked at in that light, your presence is an asset to be put to good use, deployed the way you want to use it. Whether or not you appreciate it, when you show up you're sending an unspoken message to your follower: what's important to you, what you really care about, who your care about.

If you're not in the habit of spending time on the job with your followers—some leaders are not—that may not be what you'll hear when you show up, or what your followers think you're up to. There may be a push back: "Why are you checking up on us? Don't you trust us?" If you get that, do not be dismayed: consider it to be the first step in this part of your leadership process.

That reaction can occur when the followers don't know the leader that well, or where showing up represents a big departure from the usual practice of leadership. In some operations, getting out to the job can be an undertaking requiring time and effort, like a long walk, or a drive, or coming in outside of normal working hours. "The only time we see the supervisor out on the job is when there's a problem." There are still places where those doing the work view management as the enemy.

But that's not you.

If you get resistance to showing up on the job site, have faith in yourself, your followers, and the practice. You know what your motivation is: the genuine desire to see to it that your followers work safely. Over time your followers will come to understand that; resistance will give way to an appreciation of what you are doing. That will create better relationships— "It's good to see you out here." —and ultimately trust.

But it will also take time.

THE VALUE OF OBSERVING

Simply by showing up, a leader can create a Moment of High Influence and set in motion highly impactful communication about values, commitment, and safety. As powerful as that is to effectively leading followers to work safely, in terms of managing safety performance, there's even more to the process than the message "going out" to the followers. Consider the information about safety performance that's "coming in" to the leader.

In order to be able to successfully manage safety performance, a leader needs good intelligence as to what's really going on; not what the leader would like to think is going on, or what followers would like their leader to think about that, but reality.

For better or worse, it is what it is.

Information about safety performance comes to a leader in one of two forms. The first is by a report. Safety reports are commonplace. The supervisor might call a follower who's working on a job, asking for a report: "How are things going?" A follower may report an injury or near-miss. Event reports are filed: injuries, property damage, equipment damage, and incident investigation reports. The reporting system provides that information be submitted on a periodic basis: the daily production and safety report; the weekly house-keeping inspection report, the monthly equipment-inspection report, the quarterly training report, the annual safety performance summary. There are occasions where the leader requests and receives an independent report, such as by a safety audit or assessment.

In a perfect world, reports would present the leader with an accurate depiction of reality. That only requires reading the report. Reports can, but they not always do, because they all share a common element: the information is second-hand: the source is always someone other than the leader.

That simple fact explains why so often, when a significant negative event takes place—a serious injury, for example—and the details unfold, leaders are surprised—even shocked—by what they learn. But, "Everyone knew there were problems" or "things like that had happened before" or "that practice was commonplace." If problems like these had shown up in the reports, surely good leaders would have done something to fix them.

That the phenomenon of surprise is so commonly associated with negative events argues powerfully for the other means by which to gather information about what's going on: direct observation by the leader, getting the leader's "eyes on" the situation.

The intelligence that can be gained by just being there is huge...real-time information about the work itself, the means and manner of performing the work, the environment in which the work is being done, the tools and equipment used to perform the work, and exactly what followers are—or aren't—doing.

That's the essential information every leader needs to manage safety performance.

MANAGING BY WALKING AROUND

Of the leadership activities a leader can undertake, spending time where the work takes place ranks second only to Leading by Example in its value and impact on performance. But as beneficial a leadership practice as it might be, adding hours to the workday to spend more time out on the job in the midst of followers isn't practical. The solution is found in the practice called Managing by Walking Around.

Managing by Walking Around is defined as the calculated use of the time and presence of the leader. Call it MBWA for short.

In the 1980s productivity consultants Tom Peters and Robert Waterman, authors of *In Search of Excellence*, coined the term "managing by wandering around." Observing highly effective top corporate leaders in action, a leadership practice they identified as critical to their effectiveness was the time they spent out on their business's version of the shop floor. These leaders were running some of the biggest corporate names in the world: the CEO of Nordstrom department stores, who would often be found at their retail stores; the CEO of Hewlett-Packard, who often roamed the halls of the company's research lab. These were leaders who understood where the real action was to be found in their businesses—and it wasn't behind the desk in the corner office.

Peters and Waterman described the process as "wandering around," suggesting an element of randomness to the process. Random? It's hard to imagine that these successful leaders did anything without a clear purpose in mind. In practical terms, wandering meant there wasn't a scheduled appointment or meeting driving the process. "What's our customer reaction to the new line of shoes?" The CEO could read the sales reports—or watch the sales process at the counter firsthand and talk to the sales representatives directly. "They love the new colors, but we don't have enough size 10s." Time spent out where the work of the business take place is such a valuable process, you would think that every leader would practice it every day. The leader's presence sends a message to followers, and it gives an alternative route to finding out what's really going on: see for yourself. Not all leaders do, but you should make MBWA a key part of your leadership practice.

As to how to do that, let's dispense with the random nature of the process suggested by the word "wandering." Your time is too scarce to do that. As a leader, you need to be a person on a mission: hence the use of the word *calculated* in the definition: the calculated use of the time and presence of the leader. MBWA isn't aimlessly wandering, nor is it the predictable practice of making a tour through the shop every Tuesday at 9 a.m., or spending an hour a day out on the floor. When everyone knows you're coming, you get a different version of reality.

Most importantly, MBWA isn't trying to "catch them doing something wrong." You want to see things the way they really are, the better—and the worse. Understanding that to be one of your two goals for the process—the other being to show up where it makes the statement you want to make to your followers—the question becomes, "Where do I go, and when do I go there?"

Answering that question represents calculation step in the practice of MBWA.

CALCULATION

Your goal is to see to it that everyone you are responsible for goes home safe, every day. You know your presence as a leader has a beneficial impact on your followers: they see you and you see what's really going on. As far as calculating where to go, and when to go there, what better time and place than to show up in the situations where you think your followers will be most likely to get hurt? That's the point of MBWA and managing safety.

That does not preclude you from looking at things: equipment and hazards. But if you know the most likely reasons why your followers might get hurt, you know when and where to show up. That's the best way to perform the "calculated" part of MBWA, and the results that process yields can be incredibly useful to you as the leader. The scarce and valuable resource of your time and attention as a leader is focused where the action is; in that sense, it's the second solution to your toughest challenge of finding time.

To perform the calculation, all you need to do is to ask yourself this simple question: what are the most likely reasons why my followers might get hurt, doing the kind of work they perform?

There is a key word that drives that question: why. The question doesn't ask how a follower might get hurt, or how badly they might be hurt.

Here's an example: picture someone talking on a cell phone, in a hurry, not paying attention as he hurries down a stairway. He trips over a box he didn't see, one that left on the stairway; falls and suffers a broken ankle. Is the trip and fall down a flight of stairs the reason why this person broke his ankle?

No. The trip and fall explains *how* the ankle was broken. As to the reason why he tripped and fell, take your pick: distraction, inattention, in a hurry, poor house-keeping, rule not followed, risky behavior, culture, or poor leadership. Any of those reasons would explain why someone fell in this situation.

Your answers to the question, "What are the most likely reasons my crew could get hurt?" will give you a list of reasons *why*—inattention, hurrying, housekeeping—not *how*—slips, trips, back injuries—your followers would most likely get hurt.

A list of the ten most likely reasons for injury might look like this:

1. In a hurry
2. Not paying attention
3. Not recognizing hazard
4. Complacency
5. Not trained
6. Always done it that way
7. Tired
8. Poor housekeeping
9. Not wearing PPE
10. Using defective equipment

It's easy to make your list, and the reasons make the targets of opportunity for your MBWA immediately obvious; a target being a person, place, time, situation where you think one or more of those reasons might be in play. That is how your calculation is performed.

For example, if you think hurrying is a likely leading reason why someone working for you might get hurt, take a look at the rush job of the day: the one for which there is already a built-in reason to hurry. If you think complacency is a leading reason, don't go to the most hazardous activity of the day; that's when people are paying close and careful attention. Instead, show up on a routine job that's been performed by the follower assigned a hundred times before: that's when people get complacent. If you think fatigue is a problem, head out in the late afternoon on a hot summer day. If it's inexperience, look at what one of your new people is doing. No doubt you've gotten the point: once you have a working list of the most likely reasons your crew might get hurt, when and where to show up will become obvious.

Five minutes of being in the right place at the right time can be worth hours of aimless wandering. That's all the time you need to get a sense of what's going on.

Consider the power locked up if you have the right list of top 10 reasons. More than a hundred years ago an economist named Vilfredo Pareto observed that in his native country, Italy, 80% of the land was owned by 20% of the population. This 80%/20% distribution has been found in a wide array of situations—customers, equipment reliability, quality problems, and defects—and named the Pareto Principle by quality management consultant Joseph Juran.

Do you think that 20% of the potential reasons your followers might get hurt account for 80% of the injuries that actually do hurt them? The 20% might turn out to be a very conservative number. On a long list of all the possible reasons for injury, it's very likely that a very small number of reasons will give rise to the majority of all injuries. 95%/5% might be a more accurate distribution. If that's the case, that's more good news to every leader interested in improving safety performance: focusing your effort on the right short list of reasons makes managing safety performance a lot easier.

But make sure you have the right list. When it comes to improvement, we humans tend not to be all that accurate in identifying the real source of a problem. What we think is the problem—or what we want the problem to be—might not really be the problem. Remember the Case for Safety: your follower's lives are at risk. As a leader you can't afford to be wrong about these reasons and where you spend your time.

Check your list of reasons with someone in your organization who sees all the injury reports. Something you might not think of could show up as a leading reason for injuries in your outfit; alternatively, a reason you're sure is a leading cause of injuries might turn out not to be.

That said, you probably do have the right list. With list in hand, now it's time for Managing by Walking Around.

MANAGING BY WALKING AROUND

1:55 p.m., Shop Supervisor's Office

Sitting in the office, checking the clock, the Shop Supervisor realizes it's just about time for the 2:00 p.m. round through the production area. It's been this leader's standard practice for years: checking on the status of the day's work and finding out what jobs might go past quitting time is one of those good habits this leader has cultivated.

But today things are different. In the morning, the leader decided to make up a list of the top 10 reasons why members of the crew might get hurt. The list revealed something that escaped notice: the complacency of routine work was at the top of the list, followed closely by short-service employees who lacked training and experience.

Realizing that, the leader made a decision. Instead of the 2 p.m. walk down the main aisle, there was a much better place to go to advance the cause of safety: a clean up job going on in the outdoor storage area, behind the shop. Two followers had the assignment: Charlie, the senior—and the best—crew member, assisted by the newest member of the crew, a trainee named Ron.

The work involved little more than picking up trash that littered the back part of the yard, and sorting and throwing out obsolete parts. And pallets. And tires. The yard was a mess. The JSA specified basic personal protective equipment: hard hat, glasses, and gloves.

But the more the supervisor thought about looking at the clean up job, the better the idea seemed. By visiting one simple job, in a matter of minutes the leader might be able to see and check out many of the top 10 reasons for injury: not just complacency and lack of training, but noncompliance with PPE, taking shortcuts, fatigue, failure to recognize hazards, production pressure, peer pressure and taking too much risk.

The supervisor grabbed the hard hat and safety glasses and headed out of the back door to see how the clean up job was going.

That is a perfect example of how the calculation step of MBWA looks like in practice. Heading to the yard left the rest of the crew in the production area looking at their watches, wondering, "What happened to the boss?"

THE OBSERVATION STEP

One of the biggest benefits of MBWA happens the moment you show up on the job: your presence speaks volumes to your followers. But a leader's presence works in the other direction, too: you get to see what's going on. That's the observation step of MBWA.

In a classroom, when experienced supervisors are shown a picture of people at work, they pick apart the details and see more problems than anyone would have first imagined. Take the same supervisors—one at a time—back out to their job sites, show them the work done live by one of their crew members, and the result can be the opposite. Often those same leaders will see nothing wrong. It's an example of a well-known phenomenon in psychology called cognitive bias: we have a tendency to pay close attention to some things—and ignore the important details of everything else.

2:03 p.m., Back of the Yard

It took the supervisor less than five minutes to make their way to the yard. Showing up, the supervisor was immediately struck by what they saw: a yard that yesterday was a disaster was practically pristine. Ron and Charlie had done a fabulous job, earning some well-deserved praise.

But where were they?

It took some looking. Finally, the supervisor spotted them working on the far side of the yard. Bins for scrap metal, recyclables and trash had been brought in and located by the back gate; the two were finishing up the job by sorting material into the appropriate receptacles. The supervisor headed their way.

It took two steps for the leader to realize something wasn't right. A second look made the problem clear: backs turned, hard at work, neither Ron nor Charlie was wearing any of the required PPE! No hard hat, no gloves, no glasses.

Instead of praise, the two were about to hear about their problem—as soon as the leader got to their side of the yard.

Still fifty yards away, Ron and Charlie turned and saw the boss approaching. Even from that distance, the shocked expression on their faces showed the supervisor that was last place they expected their leader to show up. A minute and a half later, the leader had made the way through the equipment in the yard to come face-to-face with the two rule-breakers.

Ron and Charlie may have been shocked to see the boss doing MBWA on their job, but what the leader saw next came as no surprise: hard hat on, glasses on, gloves on, exactly as required.

If nothing else, that familiar scenario underscores the value of Managing by Walking Around: when followers know the leader is out on the job, behavior never changes for the worse. As to the observation step of MBWA, if you're the supervisor, what do you make of the situation?

In the heat of the moment, the reaction of most leaders is predictable: the two followers are in trouble. Exactly how much trouble depends on a variety of factors: past history, other problems, the leader, administrative practices. But no matter, there's an important conversation about to take place between the leader and the two followers.

But before we go there—in the next chapter—let's spend some time understanding what else is going on in this situation. In order to get the most benefit from situations like these, a leader needs to observe *and* analyze all the important details.

In the heat of the battle, that isn't what normally happens.

ANALYZING THE DETAILS

The first thing to understand about this commonplace situation is that it represents the simplest problem a supervisor will face. There are only two people involved; there are no policies in play other than those for personal protective equipment; they have the required equipment with them, they know the rules, and their behavior changed even before the supervisor begins talking to them. Those facts remove a significant number of potential moving parts from the problem. It's clear that this is a simple matter of how two people working directly for the leader chose to behave.

By comparison, any other problem can only be more complicated. But just because it's a simple case doesn't mean it's an easy one.

Digging into the details of the situation, beyond the simple fact of not wearing the proper personal protective equipment, we can see that other safety problems are involved. For example, this was a clearly a willful violation of the rules: both knew better. How do we know that? Anytime someone changes their behavior to comply with the rules upon seeing the supervisor, it indicates the follower already knows the rules. The follower just didn't think they needed to comply, at least not until the leader showed up.

That leads to the identification of another problem. The new crew member assigned to the task undoubtedly completed basic safety training and knows what the rules are. Most likely they followed the lead of their senior peer with whom they were working. That's how culture gets transmitted from one generation to the next: it doesn't take long for the new people to act just like those who have worked there for years.

It gets worse. What does the behavior of the two employees suggest about the influence of their leader? If the adage, "The best indication of the influence of a leader is how his followers behave when they're not around," is true, this is not a good reflection on the leader. In the supervisor's defense, this kind of behavior often goes on everywhere. Why did the supervisor allow housekeeping to deteriorate to the point that it took two people the better part of a day to clean up the work area? Isn't poor housekeeping one of those reasons people get hurt?

If you're keeping track, there are now at least six problems in play in what seems like a simple case: PPE not worn, rules willfully disregarded, a senior follower teaching the new follower bad habits, new follower ignoring what was learned in basic orientation, the leader's inability to have a positive influence on his crew, bad house-keeping conditions out in the yard.

What's missing? Only the biggest problem that needs to go at the top of the list!

Sure, it's just a simple cleanup job, and the rules they're not following are just those requiring PPE. Does that mean that someone couldn't get seriously injured? Of course not. There are plenty of examples to be found where something exactly like that happened. That's the logic behind the list of the most likely reasons why people get hurt, and what brought the shop supervisor's MBWA to the cleanup job. What if one of the two were seriously hurt? How would that leave the supervisor feeling—for the rest of their life?

Once again, the Case for Safety: the Case that seemed so obvious it didn't need to be discussed. Now, in the heat of the battle where it's easy to take your eye off what matters most—sending everyone home, alive and well at the end of day—The Case serves as your compass.

That a follower could have gotten hurt is always the worst thing that could happen. It should always be the first thing that goes on the observation list.

That, and the growing list of problems found in a simple scenario makes this situation both serious—and troubling. Yes, it is serious, but on the whole, the supervisor still should see the situation as a good development: there is more good news in this situation than bad.

Starting with the best news: nobody got hurt.

Next, it's a Moment of High Influence, and a big one for both the followers. Twenty-five years later, the new crew member will vividly remember the day they got in big trouble by following the lead of a senior peer, who gave assurances the boss would never show up on a job like this. It's just a cleanup job. Bad advice. "When my supervisor showed up, I was sure I would be fired on the spot."

But don't miss out on the magnitude of the Moment for the senior follower, whose shortcomings were exposed to a new peer and the boss. "What was I thinking?"

That the behavior is willful indicates the rules are understood. It's obvious, but it rules out spending money on retraining. This is not a training problem.

The supervisor has just been presented with hard evidence as to what behavior looks like when followers don't think the leader will be around. Granted, it's not the best picture, but it is reality. The supervisor has been given an opportunity to change that behavior before a follower was injured—not after. If the two aren't the worst performers on the crew, but rather a representative sample, what does their behavior suggest about the way the rest of the crew behaves when they think the boss won't be around? Is that the culture?

Back to the top: in this seemingly simple situation there is a lot to take in, good and bad.

The time to perform the analysis isn't while two followers are working at risk. Get them out of harm's way first; handle the other problems next. But before the dust settles, ask yourself "What should I learn from this situation?" There will always be a lot.

ANALYZING THE DETAILS

BAD	GOOD
✓ Could have gotten hurt	✓ Didn't get hurt
✓ No PPE worn	✓ Had PPE
✓ Willful violation	✓ Knows the rules
✓ Influence of senior worker	✓ Moment of High Influence
✓ Your influence as a leader?	✓ Data point about real performance
✓ Housekeeping conditions	✓ Area now cleaned up

Figure 5.1   Even in the simplest situations, important details are worth analyzing.

As to how to perform this analysis, we'll borrow a tool from accountants: the ledger (see Figure 5.1). Double-entry accounting is one of the oldest accounting techniques, dating back to the ancient Egyptians at the time of the Pyramids. You know the process: debits match credits; debits entered on the left-hand column and credits on the right. It's not always a perfect match, but the technique is a very simple way to pick out the important details in a situation like this. Enter an observation on one side of the ledger—like "Willful" then ask yourself if there's an offset, something positive. Often there is: "Knows the rules."

Finally, what was learned from MBWA serves as confirmation that the supervisor's decision about the target for MBWA was right on the money. Showing up on a simple clean up job was a departure from habit.

Everyone—that includes leaders—is a creature of habit. Habit can be a good thing, but not for MBWA. Better to invest a small amount of time in the kind of good thinking that took the leader out to the cleanup job rather than taking the 2 p.m. walk through the shop, where everyone was waiting for the leader to show up.

That's the power of Managing by Walking Around.

THE LAST WORD

Managing by Walking Around capitalizes on the power of the presence of the leader. Where the leader shows up makes an important statement, and the leader can find out what's really going on in the outfit. But, for better or worse, leaders tend to fall into the practices that *their* leaders follow, emulating their behavior. When the CEO makes his annual visit to the site, there's always a series of presentations in the conference room, a catered lunch, and a tour, with the route carefully selected to make the best impression on the CEO. The last thing anyone at the site wants is for the CEO to know what the place is really like.

The CEO always seems to know how to play along. When was the last time one said, "Stop the car. I want to get out and talk to those two painting that tank to see what they think about our new paint supplier"? Who better to ask about the performance of the new coating material than the painters using it? Who better to know the answer to that question than the executive who signed the contract with that supplier?

There's an irony in all this: the single biggest problem cited in just about every attitude survey ever conducted is insufficient communication from management. The easiest way to cut through this enormous problem is to get leaders and followers talking to each other on the important topics of running the business…and doing it safely.

MBWA can do exactly that, where it matters the most: on the job.

CHAPTER 6

FOLLOWING ALL THE RULES ... ALL THE TIME

What part of the word *stop* do you not understand?

—*Charlie Hale*

Time is the biggest challenge leaders face in managing safety performance: there's a fixed amount of time available to take on what seems like an unlimited supply of things that needs to be done to see to it that every follower goes home alive and well at the end of every day. Finding time is the leader's problem. Noncompliance is a problem leaders have with their followers. Fully understood, noncompliance might well be second on the list of toughest safety challenges leaders face.

As to the significance of the compliance challenge, consider the level of safety performance that would be achieved were everyone to follow all the rules all the time. Not to just tick the box, but to fully and faithfully execute each and every safety requirement—policy, procedure, program, standard. If that were to happen, an operation might not be injury free—not every injury is caused by someone's failure to comply—but the goal of zero harm would likely be within reach.

More importantly, the odds of a major event occurring that would cause significant harm—serious or fatal injury to one or more—would likely be very low. That's because in the twenty-first century most serious hazards—"life critical" as they are sometimes referred to—are well recognized and managed by procedures such as confined space entry, equipment isolation, and elevated work. If there's a serious incident, the proximate cause of the incident is likely to have been the failure to follow the rules.

Global oil and gas giant Royal Dutch Shell studied a year's worth of its accidents and concluded that 80% of the most serious ones—their standard were ones resulting in *fatalities*—would have been prevented had all the safety procedures been followed by everyone.

While this might make perfect sense, it's well worth putting this view to the test for your operation. That's easily done by evaluating your operation's injury data base: how many of your injuries would *not* have occurred were everyone involved to be in full compliance? Your injury experience is highly likely to make the case for compliance—to you.

Were that to be the case, you might be tempted to jump to the conclusion that an enforcement campaign is called for aimed at stopping all non-compliant behaviors. That would be perfectly understandable. On the other hand, it is entirely possible that you might find yourself in the camp of leaders who've given up on achieving full compliance. These leaders think, "The truth is we humans are not naturally inclined to comply and not capable of full compliance. No sense investing time and energy on something that'll never happen. Better to focus time and resources elsewhere to manage safety performance." There is certainly a degree of truth in that view.

But, as a leader responsible for managing safety, neither strategy is appropriate—or even acceptable. Most safety rules aren't your rules. They come from laws, regulations, and your leaders: you are obligated to follow those rules and enforce them. Moreover, taking the position that compliance is an impossible mission ignores the fact that there are operations that achieve a very high level of compliance.

A high level of compliance is not an unachievable goal: it's just not an easy goal for a leader to achieve. Make no mistake: the level of compliance in an operation is a direct reflection of the contribution and performance of the leaders. Compliance is not a challenge leaders can choose to ignore. But running a "reign of terror" to make a step change in compliance is neither a smart nor an effective strategy.

The objective of this chapter is to create understanding as to the nature of the challenge of compliance: what are the reasons why compliance is the great challenge that it is. On the one hand, those reasons seem obvious: compliance is a familiar problem leaders deal with on a daily basis. Every supervisor on the planet is aware of that, but most are so busy trying to enforce the rules that they can't find the time to understand why this is the case. On the other hand, seldom are all the reasons for noncompliance considered together. That analysis we will do here, and the conclusion it leads to is anything but obvious.

If safety performance is highly dependent on compliance, and managing compliance is an important part of the job of the leader, the question then becomes, what are the best practices for leaders to follow to cause followers to follow all the rules all the time?

Answering that question begins with an understanding of the function of the safety rules.

ABOUT THE RULES

Analyzing the challenge of compliance should begin with the rules themselves. Every organization has a wide array of rules, standards, policies, and procedures covering every aspect of running the operation: production, customer service, quality, financial, information reporting, human resources, and yes, safety. These requirements come from the department, site, corporation, industry, customer, government agencies. Exactly how many rules are there? For each follower, what are all the rules they must follow?

Nobody knows. There are too many rules to count.

But, the rules are written down, somewhere. That is what makes them the rules. Moreover, they are the leader's rules. The leaders ought to know what their rules are. But leaders are too busy to add up all their rules and requirements and divide them out to identify all the rules as they apply to each and every follower.

Trace every rule to its point of origin, there's bound to be a reason why each rule was written. It can be because there was a problem: for example, when the product didn't meet customer specifications but was shipped anyway. Now there's a procedure for handline non-comforming product. A procedure can create a standard way to do something. When the company was small, anyone could call up a vendor and order what they needed. Now in a large company, there's a purchasing department, an order system, and a set of requirements that must be met for purchases from vendors. Consistency also has value: complaints that arose when the supervisor gave overtime assignments to a favorite team member led to a procedure for offering overtime fair to all.

Procedures perform another set of valuable functions: they establish the best way something is to be done. That's *effectiveness*. Procedures eliminate the need to figure out a way to do something every time the situation comes up. That's *efficiency*. In turn, rules and procedures communicate the proper and correct way do something to anyone with a stake in the outcome: the supervisor knows the overtime distribution procedure, so do the followers, and the union steward.

Rules have their functions and benefits—that's why they exist—but they come with downsides, starting with the simple fact they take away freedom. Rules remove choice from the process. Before the purchasing department issued its procedures, anyone could buy anything from anyone any time they wanted. Now there's is a set of purchasing procedures; they're someone else's rules. There is no choice but to comply with *their* rules.

Then there are the safety rules.

Safety rules serve the same fundamental functions as all other rules: promoting effectiveness and efficiency, providing knowledge, and reinforcing consistency. But there is one fundamental difference between safety rules and all other rules: what caused the safety rules to be written in the first place.

If you trace every *safety* rule back to its point of origin, you'll invariably find a tragedy, small or large. Every safety rule is a reaction to an event. Warning labels on products, for example, are there because something that wasn't good happened to someone. The rest of us are now properly warned. You might be able to legitimately criticize other policies and procedures saying, "That's a dumb idea," but you really can't say that about any safety rule.

In short, every safety rule is written in blood. Their purpose is to keep that kind of event from happening again, to a follower of yours. That's what makes the safety rules different from all the other rules. In that sense, safety rules have a higher purpose: protecting human blood and treasure.

Still, choose to pursue compliance in earnest, the first thing likely to be challenged are the rules themselves. In practice the rules as written often do have significant flaws that have been ignored because full compliance was never expected. You may hear, "Nobody follows that rule" or "Nobody does it by the rule" or "It's not humanly possible to follow that rule." In practice, the rules can be confusing, weren't intended to be followed precisely as written, and may even contradict other rules.

We'll proceed on the assumption that your rules are good as written, but you should not. If there's a push back from followers on certain rules, you should see that as a good sign: it means your followers are taking the rules seriously!

NON-COMPLIANCE

As a leader, you would like to think the truth about the safety rules would solve the compliance challenge: everyone would willingly comply with every safety rule simply because no one wants to get hurt. But in practice, the benefits to the individual from compliance with the rules are not always sufficient to cause compliance.

It's tempting to think good followers follow the rules, and bad followers are the ones who don't. That perception isn't always borne out by the facts, as illustrated by a workplace tragedy that left one person dead and a second seriously injured. The two were senior employees working in a chemical plant, with combined experience of sixty years: one a technical employee; the second a maintenance technician. Both violated a life-critical safety procedure by working inside a confined space that was later found to be filled with nitrogen. The case was investigated by the United States Chemical Safety Board, who published their investigation report. It can be found on their website: Union Carbide Nitrogen Asphyxiation.

You can be assured these were conscientious workers with every intention of going home safe at the end of the day. One had served as the coordinator of the department's behavioral safety observation program. Confined-space entry was a procedure each knew as well as anyone else in the plant. Their knowledge, experience, and motivation were not sufficient for compliance with a life-critical safety procedure.

The tragedy led to their CEO to remark in town hall meeting, "If the procedures had been followed, this never would have happened." He was not the first executive to have said that, nor will he be the last.

To the point of compliance, the question is, why would two good people not follow the rules where the potential consequences could be as severe as this? The report published by the government's investigation agency was long on ways to prevent a recurrence, but short on answers to that vital question. Perhaps it was because answering the question why would have required them to peer into the minds of the two. Perhaps it was because honest answers to the question—even if speculative—don't fit with the image the investigators want to portray to the public as being objective, scientific, and fact-based.

But isn't the function of an investigation to find out what went wrong so that appropriate steps can be taken to prevent the event from happening again?

If any good is to come from this tragedy—and the many others like this where procedures were not followed—every leader responsible for the safety of others needs to understand the answer to the question: why would good people who knew the rules not follow the rules?

Leaders need to understand the answer, not just as it applies to those who suffered the harm, but to all their followers.

FOUR FUNDAMENTAL REASONS

Why don't people follow all the safety rules all the time? If you were to brainstorm a list of the reasons based on your personal experience—as a leader and a follower—you wouldn't find it the least bit difficult to come up with a long list. It might start with "not trained" and end with "don't think they'll get caught" when people don't comply. That's not the process most leaders take when diagnosing the problem: instead, it's done on an individual case by case basis, usually after there's been a compliance failure. While no two cases are ever the same, looking at failures one by one doesn't reveal the pattern of the problem.

What's required to understand the problem is to do a "root cause of root causes." Do that and it will become clear there are four separate and distinct fundamental reasons for compliance failures. Taken together, these four not only reveal the root of the problem with compliance, they suggest a strategy to successfully deal with the challenge of compliance.

REASON 1: FOLLOWERS DON'T *UNDERSTAND* THE RULES

The brilliant automotive inventor Charles Kettering once said, "Knowing something—and understanding it—are not the same thing." Kettering didn't have the safety rules in mind, but his comment was right on point: memorizing the correct answer might be enough to pass the test taken on the computer, but safely performing a critical task that involves the subject material requires a much greater knowledge of it. How does that level of knowledge transfer happen?

If you begin to delve into safety training (as we do in Chapter 16, "Investing in Training"), it becomes evident that much of what is officially labeled as training is little more than a presentation followed by one question: "Any questions?" If the subject matter is important, there might be a test. A good test can be very useful to measure knowledge, but written tests have their limits. At best a test measures what is known immediately following the class. As to what must be understood to faithfully adhere to the policy days, weeks, and months later, that is an entirely different matter.

Class over, test passed, the individual is "qualified" to perform the task. In its various forms—computer-based, annual refresher, new employee orientation—that kind of training is all too common in the twenty-first century. But it constitutes compliance in name only. The box "trained" can be checked; an audit of the records will show compliance; those who read the reports can rest assured that the training requirement has been fulfilled. That rule has been followed. As to what it tells leaders about knowledge and understanding, it's hard to say.

Fortunately, most people understand the implications of knowledge on their safety and that of others. Understanding something requires not just knowing what and when, but also how and why. Good people don't rely on the training they're given. They teach themselves: ask questions, watch others, reread the procedures, and practice. In other words, they know *and* they understand.

But what if they don't? Is it reasonable to expect that someone who does not completely understand the requirements will be able to follow all the rules all the time?

Absolutely not.

That leads to the first rule for achieving full compliance: as a condition of compliance, those expected to follow rules must understand the rules.

2: FOLLOWERS DON'T *REMEMBER* THE RULES

Once learned, knowledge must be retained. Studies of retention of what is taught and learned indicate without immediate use, new knowledge dissipates in a matter of weeks or even days.

That's why repetition is such a great teacher. When something is done routinely and repetitively, it becomes a habit, ingrained in the synapses of the brain. It's known as "drilling." The military learned this lesson ages ago, and still practices it as well as anyone. Learning in this manner may not be the most pleasant way, but for some activities like getting out of a burning building quickly and safely, there is no substitute for a drill.

As a practical matter, from the vast array of safety rules, each follower regularly uses only relatively few. Using a rule routinely produces the same effect as drill: compliance becomes a habit. Forming good habits solves a big part of the compliance problem.

But what about the rest of the procedures, the ones that aren't used regularly?

Infrequent use doesn't change the number of rules for followers to learn and understand: it increases the probably of non-compliance simply because they seldom are required to follow most rules.

Occasional use of rules means those rules don't rise to the level of habit. Using you as an example, four years ago you learned how to use a fire extinguisher. Now the computer on your desk catches fire. Can you remember which kind of fire extinguisher to use on an electrical fire? Or safely use the right extinguisher on that fire?

It's just not realistic to expect a follower to be able to do something exactly as required when they don't do it routinely. The odds are high they won't remember what to do and how to do it by the rule. By that logic, a better solution would be to exclude the person from performing critical tasks that fit this situation. That thinking has led to the conclusion that fires are better put out by those with plenty of practice firefighting.

Thus, the second rule for achieving full compliance: frequent and perfect practice makes knowledge perfect—and permanent.

3: FOLLOWERS FAIL TO *RECOGNIZE* WHEN THE RULES APPLY

Safety rules can be divided into two categories: rules that apply all the time and rules that must be followed given specific conditions and circumstances. In practice, this difference is recognized, even if not understood in this way. Its impact on compliance is huge.

There are relatively few safety rules that apply no matter what the situation. There are Plant Rules such as smoking is not allowed anywhere on the site; no matches or lighters can be brought into the plant; the use of a seat belt is required at all times in a company vehicle; a hard hat and safety glasses are required at all times.

Safety requirements such as these are relatively simple and straightforward and apply uniformly. But even these "simple" rules are not without their complexities: for example, hard hats and safety glasses aren't required in the office building, in the rest rooms, or when going to and from the gate at the start and finish of the workday. Sitting in a parked vehicle, occupants don't have to be buckled up. Every rule seems to have an exception. If nothing else, these permitted exceptions to the rules complicate training the new people.

But most safety rules apply only when certain conditions or situations arise. Explicit in the procedure are the conditions that require its application. These can be thought of as "if or when" safety rules: *if* a specific activity is going to be done or *when* certain conditions exist.

The rationale behind this approach is to limit the time and effort involved with following the rule—which can be considerable—to those situations where the rule serves a beneficial purpose. An entry permit is required for a potentially hazardous confined space, but not to enter the gate to come to work in the morning. Fall protection is required for those working above a certain elevation, but not standing on a ladder. A face shield is required for grinding, but not when sweeping up the floor in the shop.

The approach promotes efficiency—the maximum benefit at a lower cost—but sometimes at the expense of effectiveness—the increased risk of noncompliance. The potential for noncompliance increases because these "if or when" situations require those working in the situation to first recognize the need to follow the prescribed procedure. If they don't realize the situation triggers the need for compliance, the rule won't be followed.

What's required to recognize that? First, a follower must know the requirements—and understand them well enough to be able to deal with the non-routine situation. Then, they must remember the requirement.

But understanding is necessary but not sufficient for compliance. Complying with the "if or when" type of rule also requires cognition. Cognition is the ability to recognize the situations, conditions, and circumstances that call for the rule to be followed. When the hazard and the potential consequences are significant, recognition isn't left up to the individual. By way of analogy, you don't need a passport to travel until you cross a controlled border. That's where you're met by immigration control agents, who insist you show them your valid passport.

At a border crossing, recognition is not left up to those in the situation. The immigration control process assures compliance. But for many of the "if or when" safety rules, it falls to those working in the situation to recognize the conditions that require compliance with a specific safety rule. That's what went wrong in the confined space fatality described earlier in this chapter: the two failed to recognize they were working in a confined space. That happened because *they* created the confined space by wrapping plastic over a vessel opening as they were working outside the opening, inadvertently extending the confined space to include the space where they were working. Had they recognized what they were doing, likely they would have been the first to fully comply.

Ironically, not described in the investigation report was the decision made by their leaders to exempt "non-hazardous work" done by operating personnel from the independent safety permitting process. Previously, a task such as this would have required an independent assessment of potential hazards as part of the permitting process. That independent person would have served the role of the border control agent. But that level of assurance is expensive. As part of work process reengineering—cost reduction—this independent review was eliminated, leaving the hazard and "if or when" procedure recognition up to the individuals involved. If the hazard hadn't been nitrogen and the space confined, it wouldn't have been the fatal compliance failure that it became.

While using "if or when" in requirements makes perfect sense, it makes the challenge of managing compliance far more complex. These kind of safety rules are designed to be *efficient*—requiring the least amount of effort—but not *effective* in the sense of facilitating full compliance.

The third rule for achieving full compliance: in order to comply, followers must be able to recognize that the rule applies to the situation they are in.

4: FOLLOWERS DON'T *CHOOSE* TO FOLLOW THE RULES

Finally, there is the matter of choice: followers deciding to comply. Following all the rules all the time often boils down to a matter of choice at the point of execution. That's normally what leaders focus on when thinking about compliance. But there's a difference between what's necessary and what's sufficient for compliance. It is necessary that the rules are understood and recalled, and the situation recognized as calling for the rule to be followed. Those conditions are not sufficient for compliance: those involved must then choose to comply.

Choice encompasses more than just willfully following the rule—or maliciously choosing not to. Lumped in with choice are human factors like lapses and mistakes. Consider a lapse: forgetting in the moment. Do people forget? They do it all the time. Do people forget in the moment? Sure. Do they forget where they left the car keys? Routinely. Do they forget where they put the million-dollar check for winning the lottery? Rarely. Yes, people forget what they don't regularly use, but sometimes forgetting is more a matter of making an unconscious choice driven by what's important to them.

Then there's the matter of mistakes. People are far from perfect and make a lot of errors. But the more time they take and the more care they expend, the smaller the error rate. If someone who knows how to do something correctly doesn't take the necessary care to do it correctly, is that a mistake? Or at some level, is it a choice? Form completed, checked once: they decide that's enough time spent checking.

In managing compliance, as a leader you have to think about how you view lapses and mistakes. Both will happen; both can involve compliance. What are you willing to consider as acceptable? Where do you draw the line, saying, "Enough is enough." There's a world of difference between your best performer who makes a rare lapse, and your worst performer, for whom forgetting seems an everyday event.

Lapses and mistakes aside, at the heart of the challenge of choosing whether to comply is simply this: following all the rules is never the path of least resistance. Safety rules are a burden. They require followers to do more than they would otherwise do. Following rules takes more time, slows things down, calls for attention to detail, and can require investing serious thought. That is what full compliance takes, and explains the appeal of the shortcut.

As to the return on this investment of time, effort and inconvenience to comply, people are far more likely to go home alive and well at the end of the day.

Hence, the fourth rule for achieving full compliance: the choices people make about following the rules are largely determined by how they perceive consequences. See Figure 6.1.

TO FOLLOW ALL THE RULES ALL THE TIME:

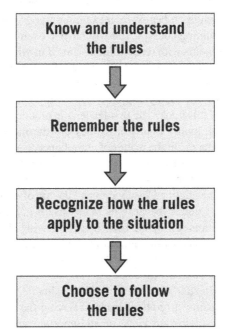

Figure 6.1 Compliance requires that four independent conditions must be met. *Choosing* becomes relevant only when the first three conditions have been satisfied.

IMPROVING COMPLIANCE

You now have an understanding as to the process of compliance that can serve as a foundation for developing a compliance improvement strategy. You understand what caused the safety rules to come into being—sadly, experience—the functions served by the rules, and how compliance benefits safety performance. Leaders instinctively see the compliance problem as a follower problem; now you appreciate the four conditions that must exist for followers to comply—understanding, remembering, recognizing, and choosing. The first three are essentially under the control of leaders. You now appreciate choosing falls to followers but is subject to the influence of leaders. There is no escaping the fact that the level of compliance is the measure of the leaders.

Armed with this knowledge, the path forward and the relevant leadership best practices start to be clear. The specifics depend on role of the leader. Supervisors routinely correct non-compliant behavior; managers have the responsibility for the design and delivery of safety training and the systems and processes to recognize "if or when"; executives play the role of articulating the case for full compliance to all their followers. Every leader needs to understand that managing compliance requires accurate information as to the state of compliance. The good news for all leaders is this requires no new rules and requirements: improving compliance involves nothing more than doing what is already expected.

But improving compliance demands a lot of work to be done by the leaders.

Taking this challenge on begins by reexamining your rules and evaluating the effectiveness of your training courses. You will have to examine the logic and schedule for re-training. Problems with recognition of the "if or when" specific rules apply may well require work process improvement and changes, much the same as has been done with improving quality using Lean Manufacturing techniques.

For the most part, the leadership practices to improve safety compliance are found in the chapters of this book. The first leadership practice is to lead by your own example. Actions speak louder than words: they are the most visible sign of your commitment to compliance. On the matter of compliance, nobody should be better at following the rules than the leader. When you do, there is tremendous leadership leverage to be gained by your example.

Dwight Eisenhower defined leadership as "the art of getting someone else to do something you want done...because he wants to do it." Influencing the choice made by followers requires selling the benefits of following all the rules all the time and the consequences when the rules aren't followed. Leaders—good leaders—know they're in the selling game. How do you sell the people you lead on the idea that they should follow all the rules all the time?

On that point, a leader's words are highly valuable: every follower always wants better communication from their leaders. Having the right words in the form of a Stump Speech to explain your views on an important subject like compliance is essential. Why should people comply with all the safety rules all the time? What do you consider full compliance? What's the best way to comply? What to do when you can't? As the leader, what are you prepared to do to help your followers to comply?

Having a Stump Speech on full compliance implies you've invested time thinking about the subject. Do that, and you'll avoid statements don't hold up under examination, like "If the procedure had been followed, this never would have happened." Safety procedures come from things like this that have happened; they are designed to prevent things like this from happening! Telling followers that compliance would have prevented this adds no value to the conversation.

How do you know the level of compliance of your followers? Chapter 5, "Managing by Walking Around" explains the best practice to find out. Chapter 17, "Measuring Safety Performance" will give you measurement techniques to look not just at safety results but also into safety processes, including compliance.

Comes the time to deal with non-compliant behavior, leadership best practices to correct behavior are found in Chapter 8, "Behavior, Consequences—and Attitude." Dealing with more pervasive behavior is explained in Chapter 12, "Managing Accountability." The benefits from recognizing and reinforcing compliant behavior are also described in this chapter. If you're a manager or executive looking at creating a culture of compliance, you'll find useful information in Chapter 15, "Creating the Culture You Want." Improving the effectiveness of training so that it creates understanding is a management undertaking. The ways and means to accomplish that goal as it relates to safety training are described in Chapter 16, "Investing in Training."

Finally, were every follower to follow all the rules all the time, the life of a leader—from the front-line to the executive suite—would be so much easier. Think of all the problems you wouldn't have to deal with.

CHAPTER 7

RECOGNIZING HAZARDS AND MANAGING RISK

As ever, Watson, you see but do not observe.

—Sherlock Holmes

In the preceding chapter, "Following All the Rules … All the Time," we examined the challenge of following the rules. To achieve compliance a leader faces a brutally tough set of challenges that will never end. The fact that everyone follows all the rules today doesn't guarantee everyone will do that again tomorrow. Or immediately comply with the new policy rolled out in the morning safety meeting.

But what if you were able to accomplish the goal of full compliance: everyone under your supervision followed every safety rule every working day? And they didn't just "tick the boxes," but fully and faithfully followed both the letter and spirit of the requirements. If that happened, safety performance undoubtedly would be fabulous. But would following all the safety rules all the time guarantee that nobody would ever get hurt?

Unfortunately, the answer is no. There are plenty of ways to get hurt that don't involve breaking the rules. If your goal is to see to it that every follower goes home safe every single day, your challenge goes beyond following all the rules all the time. You face a second challenge that might be the equal of the challenge of full compliance: you and your followers must be able to *recognize* what can hurt *them*. That recognition needs to be done in real time and in the dynamic world of your operation. And as a result of recognizing what can hurt them, you and your followers must take preventive action *before* someone gets hurt—not after.

Successfully dealing with the challenge of recognizing hazards and in turn, taking the measures to reduce risk requires understanding, awareness and a practical set of tools. Understanding begins with knowing the answers to these three fundamental questions about hazards:

- *What* is a hazard?
- *Where* do I look for hazards?
- *How* do I look for hazards?

Alive and Well at the End of the Day: The Supervisor's Guide to Managing Safety in Operations, Second Edition. Paul D. Balmert.

Understanding the answers to what, where and how leads to other important questions about risk:

- Is a hazard the same as a risk?
- Can my operation be run with zero risk?
- How do I manage and reduce risk?

For safety, your role as the leader can be looked at as "managing risk." You do that every day, and the work starts with recognizing hazards.

WHAT IS A HAZARD?

Being able to recognize a hazard starts with knowing what a hazard is. The definition is simple enough: a hazard is a source of danger, something with the potential to harm someone. Simple and understandable as this definition is, it sets up the first big challenge in recognizing hazards: if a hazard is something that can hurt someone, what is *not* a hazard?

Ask followers what the hazards are for their jobs, their answers typically start with those most serious: hazardous materials, operating equipment, high temperatures, high voltage electrical equipment, working at height. But those answers seldom match up with the things that get most people hurt. It's usually mundane, ordinary things found in the workplace that do the damage: stairways, ladders, floors, hand tools, drums, bags, and pickup trucks.

If you were to create an inventory of all the things capable of hurting your followers as they go about doing their jobs, it would be a very long and continually growing list. Can you think of something that isn't hazardous in some situation? Water? There's ice, steam, condensate, high pressure water, and always the possibility of drowning. A computer on a desktop? A few decades ago, there was no such thing; now carpal tunnel syndrome from excessive use of the keyboard has become commonplace. Under the wrong set of circumstances, just about anything is capable of producing harm, and even serious harm. A trip and fall down a flight of stairs can—and has been—fatal. If your goal is to see to it that everyone goes home safe, the number of things you have to worry about—and manage—can become so mesmerizing you don't even know where to start.

As a first step, let's make sense of what looks like a random list of hazards. If you work from the injury backwards to determine the hazard—the source of danger, you will find there are three separate and distinct elements that combine to produce the harm:

- There was a <u>person</u> suffering the injury.
- There was an <u>object</u> doing the harm to the person.
- <u>Energy</u> was required to make the object harmful and bring the three elements together at the same time and in the same place.

The combination of these three elements is what it takes for someone to get hurt. You can find a few exceptions, but those are rare, meaning these three elements create very useful model to explain injury causation, or how people get hurt.

THE INJURY TRIANGLE

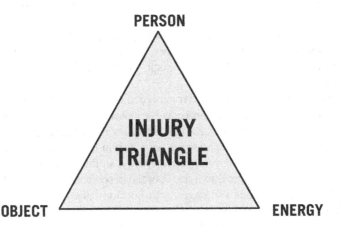

Figure 7.1 An injury requires Person, Object and Energy coming together at the same time and place.

Taken together, these three elements form what can be thought of as the Injury Triangle (see Figure 7.1).

Remove the person, it's impossible for anyone to get hurt. The base of the Injury Triangle creates the hazard that does the harm: an object with enough energy to make it harmful to a person. By this model, anything can become a Hazard: reading reports of all the different things that have injured people confirms the validity of that part of the model.

In practice, the Injury Triangle is a dynamic model. Every person is continuously creating a changing set of hazards as they move through the day: where they are and what they are doing at any point of time puts different things and energy within proximity of them. There is a set of hazards in the room where the safety meeting is held at the start of the shift; there is a set of hazards created in transit to the job site; a set of hazards created on the job, as work begins. In a real way, the only constant in the Injury Triangle is the Person in the model. Each person is constantly creating a set of hazards that can harm them, making hazard recognition a continuous (rather than static) process.

The inherent properties of an Object can set up a hazard: for example, a toxic raw material used to manufacture a product. But there is more to creating a hazard than the properties of the Object. The interrelationship among the three elements also serves to create a hazard; that's the way Energy fits into the model. For example, an object is first lifted, and later dropped; something is heated, and then touched. The person can contribute the energy to make an object harmful: descending a stairway sets up a trip and fall; repetitive motion causes a soft tissue injury.

Everywhere you look, you'll see the elements making up the Injury Triangle, because it takes people, objects, and energy to do the work of the business. Simply by virtue of placing people in a workplace with objects and energy, hazards are created. In theory, People could be removed from the Injury Triangle by insisting everyone stay home; that would guarantee the operation would be injury free, but that's no solution because nothing would get done.

Granted, some objects are far more inherently dangerous than others: methyl isocyanate, which did so much tragic harm in the industrial accident at Bhopal in 1984 was one of the most deadly products ever manufactured. But one of the great necessities of life, water has claimed many, many lives. So have things intended to make the job safer like ladders, scaffolds and stairways. The truth about hazards is that, under the wrong set of circumstances, just about any object can prove fatal. Do not make the mistake of labeling some hazards as life critical and the rest nonhazardous.

It takes all three elements—person, object, energy—to produce an injury; the removal of any one of the three prevents the possibility of an injury. In that way safety and loss control procedures work at "breaking the triangle." The foundation of many safety procedures is to separate the person from the object, in time and space, so that there can be no injury; remove the energy, so there can be no consequence from contact; inspect the object, so it does not fail in service. An electrician is permitted to work on high voltage switchgear, but not normally when it is energized. A pipefitter is permitted to break into a line, but only after the contents and pressure have been removed. The lockout procedure ensures the removal of the hazardous energy before any work can start. The confined space procedure ensures the removal of the dangerous object—whether that is hazardous material or the absence of a breathable atmosphere—before entry. Equipment is designed, inspected, maintained, and certified to prevent failure in service.

The design principle for procedures like these is keeping one of the three elements away from the other two. That breaks the Injury Triangle.

A different set of procedures provides benefit at the point when the three elements of the Injury Triangle have been set in motion and may come together at the same time and place. These procedures function like a safety net, there to provide protection when needed. PPE is the best example.

In terms of the Injury Triangle, personal protective equipment provides a barrier between the person and either object or energy. Seat belts keep the driver from making contact with the windshield of the vehicle in the event of a sudden stop or collision. A hard hat spreads out the energy from the dropped object, lessening the force per square inch. A fall arrest device prevents contact with the ground in the event of a fall.

Personal protective equipment has its benefits and can prevent harm to the person. Its protection has its limits—but it provides no benefit at all if it's not used.

Like a safety net, this additional level of protection isn't needed *until* the three elements of the Injury Triangle are in play. Since most of the time a safety net isn't required, it's easy to understand why requirements involving personal safety equipment are often the first safety rules broken. A safety net also provides no benefit when it's not needed: unless there is an event, there is no harm to the person who's not wearing their PPE.

The problem is that safety nets can't instantaneously be strung up when the hazard materializes. That's why PPE must be routinely and completely used: you don't know when you'll actually need it.

A former industrial painter named Russell serves as a tragic illustration. Russell became a motivational speaker after he suffered a fall at work. Now confined to a wheelchair, in a very plainspoken way, Russell talks about what it was like to fall 30 feet and survive the experience. His is a very powerful message. There is one small footnote in Russell's account that was striking, and relevant to the point of the safety net: that's what the attending Emergency Medical Technician had to say when he arrived on the scene. Russell was conscious, the first thing the EMT said was, "You're wearing a harness—so why weren't you tied off?"

It's a question Russell must have asked himself hundreds of times.

Fortunately, it's relatively rare for person, object and energy to come together to produce an injury, but not uncommon for all three elements to be in play and not to meet up at the same time and in the same place. That's what is known as a near-miss (or near-hit or close call) as for example, a dropped scaffold board that barely misses someone working on the ground below. Whether a miss or a hit, an event occurred and all three elements of the Injury Triangle were in play: Person, Object and Energy, with one of the three serving as the trigger.

As to that trigger—the *cause* of the event—it is the same whether it produced a hit or a miss. Whether the scaffold board hits the ground and is later picked up, or it hits and fatally injures the person working below, the cause of the event would be exactly the same.

That means the difference between a hit and a miss is found in its *effect*. In the example of the dropped scaffold board, it's a life and death difference.

WHERE DO I LOOK FOR HAZARDS?

The answer to what defines or creates a hazard is simple: the Injury Triangle. The processes involved in looking for hazards—recognizing—are complex and sophisticated. In the twenty-first century, industrial operations the world over have a robust set of policies, procedures, programs and processes designed to both identify and manage hazards.

Using a capital investment in a major piece of production equipment as an example, the hazard recognition process might begin with a design safety review. When construction is completed, there might be a pre-startup safety review. New raw materials would be accompanied by Material Safety Data Sheets. Standard Operating Procedures would be written, and operating staff trained. The equipment would be periodically inspected. Specific tasks would be done following safety procedures like confined space entry, equipment isolation, work permits and job safety analysis. Industrial hygiene monitoring might be performed to identify hazards found in the environment as the work is done.

Taken together, these individual elements form a web of protection, serving to first identify and then to manage hazards associated with operation. A system such as this is substantial and very impressive. It requires an investment of time, energy and money. The question every leader—and follower—must ask is this: does this web of protection guarantee that every hazard will be identified and managed?

The answer is no; the system is not the solution. As vital as this system is to managing safety performance, every person working in the operation still must be on the lookout for what can hurt *them*.

Where, then, to look for those hazards?

At the level of the individual, the hazards that can harm any person come from three separate and distinct sources.

- The first route is via the specific **task** the person is assigned to perform. An auto-repair technician is removing the lug nuts on the front wheels. The air wrench used to do that task slips and hits his face. The injury report reads, "Impact wrench chips tooth."

- The second route is via the larger **environment** in which the job is performed. Work is always performed in a three-dimensional space; there are surroundings. There not be a thing unsafe in the task—the person is performing it exactly as prescribed—yet something unrelated—above, below, around—causes the injury. The injury report reads: "Passing motorist strikes road worker."

- There is a third route: situations where there is nothing "hazardous" about either the task or the environment. The injury report reads, "Lifting scaffold equipment causes back spasms." Something happened between the person and the task performed—the amount of lifting, the position from which the material is lifted, the physical condition of the person performing the task—causing the injury. The cause or source might be labeled as ergonomics, but there are cases where the person says, "My back started hurting when I bent over to pick up the wrench" or "I re-injured my sore knee." The source of these types of injuries is best considered as coming from **Inside** the person.

These three routes of hazards explain where to look for hazards that can harm the individual. They can be remembered as **TIE: Task, Inside** the person, and from the **Environment.** Each route is entirely independent of the other two; a hazard found in any source can be more than capable of sending someone home hurt at the end of the day. Hazards from each source need to be recognized and managed.

Does the existing system your operation uses—your web of processes—guarantee that hazards from *all* three of these sources will be identified and managed?

Likely not.

If your goal is that every one of your followers goes home safe every single day, it is mandatory to defend them—and help them defend themselves—against all three routes of "incoming fire from the enemy."

For that, you and your followers have **Three Lines of Defense**.

THREE LINES OF DEFENSE

The first line of defense—being able to recognize hazards—is knowledge: what you don't know *can* hurt you. Plenty of people have been hurt trying to do something they didn't know how to do. There will always be more hazards than there are procedures to control them, making leaders dependent on the knowledge of their followers to keep them safe. So are followers.

Knowledge as the primary defense puts training front and center. If the programs for safety training—as well as teaching job skills and knowledge—aren't effective, followers will be left defenseless against injury. That may be obvious, but so is the fact that many training programs are far from effective. Coupled with new people coming into the industrial workforce the need for effective training becomes more acute. That challenge is treated in Chapter 16, "Investing in Training." Training and qualification are a vitally important part of seeing to it that everyone goes home safe.

But let's make the assumption all of your followers actually know what they're doing out on the job. Most do. Is their knowledge sufficient for each to recognize—and take the necessary safeguards against—the hazards capable of causing injury on the specific jobs they are assigned?

No. Knowledge may be necessary to work safely, but it is far from sufficient. If it were possible to go back to the moment prior to an injury—leaning on a ladder before it fell, gripping the wrench before it slipped, stepping out of the car before slipping on the ice—and ask the person who got hurt, "Do you think you could get hurt doing that?" in the most cases the answer would be, "Yes, I suppose something like that could happen. But it won't. At least not to me, right here, right now."

So much for, "Trust me; I know what I'm doing."

That people can—and often do—get hurt doing things they know how to do safely is a principal reason for the second line of defense against hazards at the point of execution, where the individual performs the task. Knowing that training is not sufficient to keep people safe, every organization employs an array of policies, procedures, standard operating procedures, and safe work practices to recognize and manage hazards. These range from broad requirements, like the policy that mandates minimum job training and qualification, site rules like requirements for personal protective equipment, to activity-specific safety procedures such as those for safe work permits, job safety analysis, excavation, confined space entry, working at heights, and hot work. All are part of the system, weaving a net of protection against hazards, first by identifying the hazard, and then by prescribing the appropriate defense.

But is your hazard recognition system the solution? Is it enough for a follower to think "I'm following all the procedures, so I can't get hurt"? Relying on the system of safety procedures to keep your followers safe seems like an effective approach—until you delve into it. Then the flaws in thinking that become obvious.

It's fair to say that every safety procedure came about because something unexpected happened. That still happens. You never want to learn something the hard way and have a new procedure created and named after one of your followers.

As you know, procedures aren't always complied with. Even when a follower is following the rules as prescribed is no guarantee they can't be hurt by someone else who didn't follow the rules prescribed for them.

The "if or when" safety procedures require that people must recognize the conditions requiring that the procedures be followed. That is something that doesn't happen automatically. If someone fails to recognize that conditions call for additional precautions, their lack of compliance isn't making the wrong choice, but rather a failure of what's known as *cognition:* the combination of knowing something, remembering it, and then recognizing when it applies in the moment. *Recognition* is the more practical description (see Figure 7.2).

COMPLIANCE REQUIRES HAZARD RECOGNITION

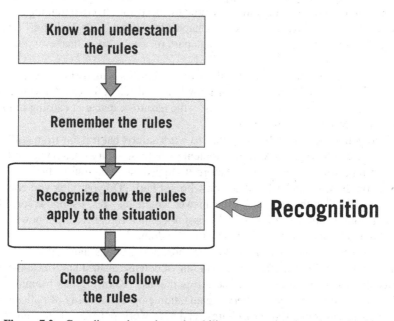

Figure 7.2   Compliance depends on the ability to recognize that a hazard exists.

Finally, there are ways to go home hurt or hurting that don't involve breaking the safety rules. These are often found in the second and third routes of "incoming fire":

Inside: the relationship between the task and the person assigned the task: "Lifting scaffold equipment causes back spasms."

Environment in which the job is performed: "Passing motorist strikes road worker."

Hazards from these sources arise from the specific relationship between the person and the task, or from a hazard not directly associated with the task assigned. Situations like these might be foreseeable, and procedures like a job safety analysis could identify the potential hazard. But these procedures don't function that way because of design or application: they don't ask questions about the specific person assigned to the task or objects unrelated to the task. In other words, they leave a gap.

If your procedures aren't capable of identifying and managing these second and third routes, you don't want to find that out after a follower has been hurt. There is a simple way to perform a "leak check" to see if your procedures are watertight. That is to construct a matrix with the three routes on one axis and the relevant safety procedures on the other. This exercise will tell you how well—if at all—specific safety procedures like work authorization, confined space, trenching and excavation, and even the job safety analysis your operation uses do to identify the potential hazards from each route. If you are confident your procedures provide complete

THE SYSTEM IS NOT THE SOLUTION

Policy or Procedure	Assigned Task (e.g.: replace gasket)	Inside (e.g.: body position)	Environment (e.g.: work overhead)
Work Authorization Policy	X	?	X
Confined Space Entry	X	?	X
Trenching and Excavation	X	X	?

Figure 7.3   Before trusting a hazard identification procedure will identify all potential hazards, the better practice is to verify that it will.

identification, you can then rely on the procedure to do the heavy lifting of hazard recognition for you.

What you're more likely to find is, as valuable as these procedures are, they won't guarantee that every source of danger will be identified. (See Figure 7.3).

HOW DO I RECOGNIZE HAZARDS IN REAL TIME?

As important as knowledge and procedures are, going home alive and well at the end of every day demands the capacity to recognize hazards in the real world and real time of work. The dynamic model of the Injury Triangle suggests the approach to creating the Third Line of Defense: the ability on the part of your followers to recognize where people, objects, and energy have a significant potential to combine to produce an injury (See Figure 7.4).

Since the three elements come together only when an event occurs, on the job hazard recognition requires observing what's present and then processing that information to form an opinion as to the potential sources of danger. The deductive abilities of the brilliant detective Sherlock Holmes kept his sidekick, Dr. Watson, in a constant state of amazement. Holmes would explain all the seemingly minor details that led him to a startling conclusion, and then put a finish on his explanation: "As ever, you see, but you do not observe."

Differences in the capability to recognize hazards on the job are created by the confluence of many factors: training, education, job experience, life experience, even interests. If you show a picture of a specific work activity to a range of people—for

THREE LINES OF DEFENSE

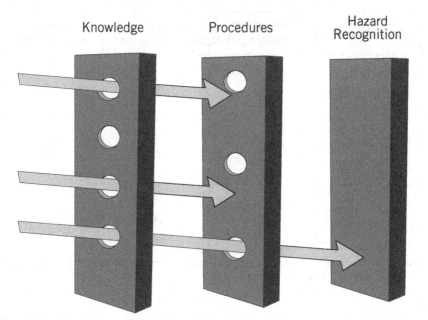

Knowledge   Procedures   Hazard Recognition

Figure 7.4 The third—and last—Line of Defense is the ability to recognize hazards in real time.

example, an experienced operator in the department where the photo was taken, an experienced maintenance technician, a design engineer, and a brand-new trainee—and ask them to name the hazards they see, their answers will vary widely. Each will have their own view of the hazards, some similar, but some significantly different. The operator would know about the hazards associated with the raw material; the maintenance technician would know of the hazards that can come from poor lubrication; the designer would know about materials of construction. A brand-new trainee would be the one most likely to recognize the most obvious hazard—poor housekeeping in the department—others had stopped seeing because they've all been there for years.

That demonstrates a key principle in process of recognizing hazards: two sets of eyes are far better than one. Given different knowledge, skills and experiences, hazard recognition improves by the involvement of more people. Safety is the ultimate team game. But ultimately the person doing the work has the most to gain or lose from recognizing what can get him or her hurt.

In addition to a second set of eyes, here are five other ways to better recognize hazards:

1. Factor in the physical **environment.** It isn't just the task being performed that represents the hazard; it's where the task is being done. Evaluate the larger

environment that envelops the task: what's above, around, next to, what others are doing, and most assuredly the weather.

2. Put your **head on a swivel.** That's what you do when in a hostile environment. Focus on the big picture *and* the details, looking for the unexpected. Look in every direction: left, right, up and down, ahead, behind.

3. Pay attention to what's **changing.** Work is dynamic, not static; don't assume otherwise. The specific work to be done can easily change as the job progresses, leading to unanticipated hazards. Even if the task doesn't change, the people doing it can, and so can factors in the environment.

4. Use all your **senses.** As a source of danger, a hazard has physical characteristics. The five senses—sight, sound, smell, touch, and taste—can be the best set of hazard identification tools anyone has. Then there is what's known as the sixth sense: intuition. In the case of recognizing hazards, feeling that sixth sense might be considered *foreboding*—the sense that something isn't right, but what's wrong can't immediately be quantified. The human brain is a magnificent collector and processor of information, far more powerful than we give it credit for being. If something doesn't seem right, it's probably because something *isn't* right. Take that as a cue to stop and determine exactly what that something is.

5. Look for the **Warning Flags.** They include any signal that indicates the risk of an injury is rising. Consciously or subconsciously, over time supervisors develop their own list of signs of trouble, some obvious, some subtle. Here is a list of potential candidates, several of which build on what has previously been described:

- Changing work scope
- Changing/inclement environment
- Someone in a hurry
- Improvised tools or work methods
- Unusual smell, sound, appearance, touch, taste
- Elevated/overhead work
- Working around hazardous material/energy
- Awkward/stressed work posture
- Not following the procedure
- Poor housekeeping

You are now equipped to better perform and manage hazard recognition at the point where the work is done. The **Injury Triangle** frames what you need to pay attention to: the *objects* with sufficient *energy* to pose the potential for harm to your *people*. **TIE** tells you where to look for those hazards: *task, inside, and environment*. **Warning Flags** suggest what to look for: indications of where there is the significant potential for harm.

WHAT IS RISK?

As important as hazard recognition is to working safely, there's an even bigger challenge lurking below the surface: managing risk.

The terms hazard and risk are often used interchangeably. For example, a "hazard assessment" in one operation might be called as "risk analysis" in another. "Risk mitigation" might mean exactly the same thing as "hazard control" or it could involve something entirely different, like requiring additional PPE. Leaders running the business are free to call any policy by any name they choose, and operationally define any term the way they want to. But no matter the names chosen or how they're used, leaders can't change the definition and meaning and significance of these words as they exist in a language. In managing safety, there is a world of difference between the words *hazard* and *risk*.

Sadly, that difference is poorly understood by industrial leaders from the frontline supervisor to the CEO. The experts themselves frequently confuse the two. A proper understanding of each—hazard and risk—won't make life easier for the leader (it will make it tougher) but it can make safety performance better. A proper understanding can spell the difference between life and death.

So, let's begin with operational definitions of each:

- A *hazard* is a source of danger; something that can cause harm. In the Injury Triangle, Object and Energy combine to create that source.
- A *risk* is a measure of probability; the odds a hazard will cause harm.

That is not the way many leaders define the terms, particularly risk. Consider the implications of these two definitions:

Hazards are real; they exist in real life. Hazards are things that can hurt people.

Risk is an abstract concept. It's a calculation: a number written on a piece of paper or white board.

As a measure of probability, risk does not guarantee what will happen. If something will happen, it is not a risk: it is a guaranteed outcome. If something cannot happen, there is no risk.

Different as they are, hazard and risk are inseparable. For every hazard there is a risk. It is as if they are the head and tail of the same coin: the hazard is the source of danger; the risk is the probability the hazard causes harm.

When something happens, it creates an event. In the model of the Injury Triangle, an event means Person, Object and Energy are in play. If they come together at the same time and in the same place, there is a hit. A hit may cause harm. If the three do not come together, it is a miss (or *near-hit*).

Consequences are determined when the event happens. For safety, consequences range from no harm to fatal harm. Seemingly minor hazards have proven fatal; people have survived events involving life-critical hazards (Figure 7.5).

Ironically, the word risk can be traced to its Arabic origin: *rizk*. *Rizk* is what each person is given in their life: things like family, talent, wealth. No matter what

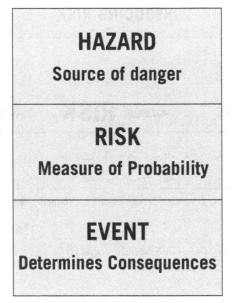

Figure 7.5   For every hazard there is a risk. Seemingly minor hazards have caused serious harm.

they might be for any individual, the potential consequences from an event puts them all in play.

Managing risk is part of living. With a pandemic, we learned about the hazard of COVID. Shortly thereafter, we were informed about the range of consequences were we to contract COVID: from no symptoms to the possibility that COVID could prove fatal. Of course, we won't know our consequences until we test positive; then we find out.

Isolation, sanitation, masks, temperature checks and social distance were the means to reduce risk: to lessen the probability of contacting COVID. The vaccine was initially promised as a cure: get the shot, and you won't catch COVID. Later we learned the vaccine didn't eliminate the hazard; it might not change the probability of contracting COVID; it does appear to have mitigated the consequences. With the shot, COVID was less like to prove fatal or as severe.

Risk describes uncertainty. We routinely make choices without being certain about the outcome. Do I wear a mask? Do I get a flu shot? Do I buy flight or trip insurance for my vacation?

In addition to the hazard, time, effort, cost, and convenience factor into our decisions.

ZERO RISK

The Injury Triangle suggests that whenever there are people present, there will be things that can harm them. That raises the question of zero risk. There is a long list of hazards found in your operation. Your goal is to send everyone home alive and

REDUCING RISK

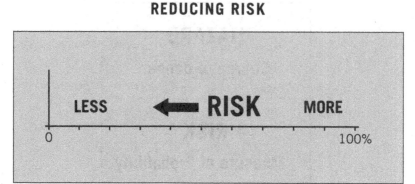

Figure 7.6 Achieving Zero Risk is impossible, but reducing risk is readily achievable.

well at the end of every day. Understanding that to be so leads to this question: can you run your operation with zero risk?

Of course you can't.

Guaranteeing that no possible harm could happen would require removing either all the hazards or all the people, neither of which is possible. As a practical matter, if you're planning to stay in business, your goal can't be to achieve zero risk. It should be to reduce risk to what you consider an acceptable level (Figure 7.6).

Zero Risk is a direction—not a destination. You can't eliminate all risk, but you can always do things to reduce risk. If you're concerned about the risk of someone falling off a ladder, you can provide a scissor lift or build a scaffold. But still, people have been known to fall from both.

What you never want to do is to increase risk: making the probability of an event higher. Working from ladder with only one point of contact increases the odds of falling. If you think that's a violation of a safety rule, consider this: every time anyone does not comply with a safety rule, they're increasing risk. Another point on behalf of compliance.

Hazards are real; when an event happens, the consequences are real. They're what happen to people. The severity of consequences can be managed. For example, a person working on a ladder could be required to wear fall protection. If they do, that won't change the risk—probability—of falling off the ladder, but it will lessen the consequences if someone were to fall.

Doing that is called mitigation: lessening the effect of an event. As an example, PPE mitigates the consequences when an event happens. Personal protective equipment does not change the probability of having an event, but it might prevent someone from being harmed by the event.

So, consequences are mitigated; risk is reduced; a hazard is eliminated.

Now that you understand the difference between hazard and risk, the next time you read the finding on an injury investigation report, "Failed to recognize the hazard" you will know to ask, was this a case of hazard recognition—"I didn't *see* the hazard" —or was it a matter of misjudging risk—"I didn't *think* that was going happen to me."?

That said, it is entirely possible for someone to miss the obvious and not even realize a hazard is there in the first place. We humans can fall victim to what is called Cognitive Bias: the tricks our brains play on us to make life easier.

Easier is not always safer.

WHAT IS THE RISK?

When it comes to recognizing—and managing—hazards, there are always two questions to be answered. First, what are the hazards, the sources of danger? Second, what are the risks, the likelihood that any hazard will occur? For every hazard there is a risk.

For any task, such as changing a tire on a car, there are various hazards with the potential to cause someone to get hurt. Those hazards are largely known or knowable: rarely does someone get hurt from a hazard nobody knew existed. In the case of hazards there are ample authoritative sources of information from which to know about them: training, formal education, experience, operating manuals, product literature, warning labels, and people who share what they've learned, sometimes the hard way.

By comparison, determining the risk for a hazard means estimating the probability of an event occurring. Intuitively we know the probability of hazards varies widely: the auto mechanic using an impact wrench to change the tires knows suffering hearing loss from noise exposure is more likely than succumbing to carbon monoxide poisoning from the vehicle's exhaust system. Both could happen; one is far worse; the other a lot more likely. How much more likely?

Where do you go for authoritative information to find out the risk for these two specific hazards?

For some life activities, a great deal of statistical information about risk has been collected. In the matter of private automobiles, you can look up the probability of having your vehicle struck on the road by another vehicle, hitting a pedestrian while driving, or being killed in a vehicle accident. But you can't find the odds of burning your arm while changing the spark plugs. If you're supervising an auto repair shop and are concerned about safety, knowing that risk would be very useful.

One of the great shortcomings in industrial safety is that there is very little authoritative information about the risk of injury from the hazards found on the job. We know about slips and falls, and there are studies of the impact of falls from varying heights on the human body, which have been used to determine the level at which fall protection is required. But we really don't have a study telling us the probability of falling down a flight of stairs, or falling while working at a given height, such as five feet. One reason is that performing that calculation requires determining exposure: how often do people actually go up and down stairs, or find themselves working in places where they might fall? Remember, risk is simply a matter of probability.

Climbing stairs is but one small task out of the hundreds that one person does every day. When you think about it in those terms you begin to appreciate the magnitude of the process of calculating risk for job hazards: it's just too complex an undertaking. There are too many hazards and too many variables. Because it's that complex, we don't have hard numbers about industrial risks. Because we don't have hard numbers telling us what the risk is, we rely on our sense of what seems risky.

There's good news and bad news. The good news: when we have an elevated sense of risk we act carefully. By acting carefully, the odds of the hazard harming us decreases. The bad news: just because we don't sense risk does not mean there is no hazard or no probability of harm from a hazard. When we're convinced there is little to no risk, we act less carefully. When we act less carefully, the odds of being harmed by the hazard, increase.

That's the **Law of Risk.** We can – and do – alter risk.

ASSESSING RISK

Recognizing hazards is a matter of physical science: seeing what's there. In a perfect world, experts would tell us the risk of harm from every hazard we encounter. That way, we'd go about managing risk in the most intelligent way. We'd be careful where that would do us the most benefit. In real life, we're left to calculate risk on our own, figuring the odds that something might go wrong.

What we know about how we calculate the odds is this: we aren't all that good at calculating the correct probabilities for the occurrence of hazards. We rely on our sense of risk. The research conducted on risk perception confirms that finding. How we go about sensing risk is rooted more in human psychology than it is in statistical science. We worry about hazards that aren't very likely to hurt us, and don't pay nearly enough attention to the things that most likely will.

There are plenty of everyday examples that bear out this conclusion. Commercial aviation is by far the safest means of transportation. But after a turbulent flight and a bumpy landing, a very nervous passenger will jump into their car and speed home, without buckling up, and all the while talking on the cell phone. The first bolt of lightning will close down the golf course, but fair-skinned golfers don't think to apply sunscreen. A driver who routinely speeds will still carefully obey the requirement to stop for a school bus discharging small children.

In each case, the hazard that is greatly feared—an airline crash, being struck by lightning, hitting an innocent child—is far less likely to occur than the hazard that isn't. Research by communications professor Peter Sandman provides a useful foundation for understanding how people view, and make decisions about risks to their safety and health. His conclusion: the hazards that "scare people and…kill people are not the same."

In Dr. Sandman's view there is a certain logic to this thinking— but not logic based on hazard and probability. When a hazard is catastrophic, memorable, dreaded—and under the control of someone else—it is perceived as very risky. With that risk comes fear, in some cases a healthy and productive fear. One reason commercial aviation is so safe is that, if it were otherwise, precious few of us would choose to fly. On the other hand, when the hazard is common, everyday, ordinary, under our control and the potential consequences chronic or incremental, there is far less fear. We become complacent.

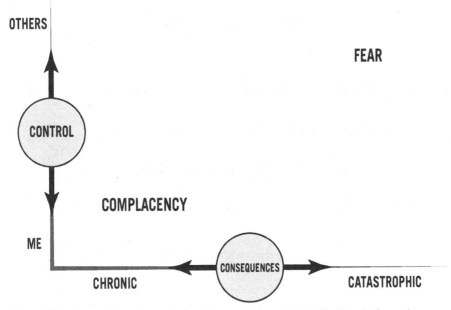

Figure 7.7   Any task can be mapped to the matrix to predict the likelihood of complacency.

This theory offers the best explanation of why traffic always comes to a screeching halt when the school bus stops: failure is catastrophic, memorable, and dreaded. It is relatively easy to perceive those kinds of consequences, and no one would want to be responsible for causing that kind of tragedy.

Sandman's view goes a long way toward explaining more than just why we tend to be nervous airline passengers and fear lightning. It points to the situations when we are very careful and those when we are likely to become complacent. That view suggests a way to influence how followers perceive and manage risk on their jobs.

The two variables of control and appearance of consequences combine to predict the degree of "fear versus complacency" a follower will sense:

- High control and chronic effect: complacency
- Low control and catastrophic effect: fear

The combination of these two factors produces a matrix that can be used to predict where complacency is likely to be found. (See Figure 7.7) Most injuries will be found in the lower left quadrant, where most people spend most of their time working.

FOUR RULES FOR REDUCING RISK

Risk perception seldom matches reality. When it comes to recognizing what can get someone hurt and taking the necessary precautions to reduce the odds that it will, you're always fighting an uphill battle. You are called upon to influence your

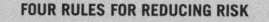

FOUR RULES FOR REDUCING RISK

1. To reduce risk, you must first recognize risk.

2. Don't up the odds: Follow your procedures.

3. Make everyday hazards a clear and present danger.

4. Reducing risk requires people willing to say "Stop."

Figure 7.8 These rules are a practical way to manage and reduce risk.

followers to take the time and effort to first observe all the hazards present, take the ones with higher risk seriously, and then take steps to (as the circumstances dictate) remove the hazard, reduce risk, or mitigate the consequences should there be an event.

That is no small feat. Here are **Four Rules for Reducing Risk** that will help (Figure 7.8):

Rule Number 1: To Reduce Risk, You Must First Recognize Risk

Thanks to Dr. Sandman, we now know that humans discount familiar hazards. It's not that they "don't know," it's that they think, "It will never happen ... to me" or if it does it won't be bad.

Being armed with the facts is a good start. There are plenty of statistics available about what's likely to hurt someone. Your safety staff is a good place to start. They can tell you about the hazards and the probabilities for your operation and for the larger world of work. For example, statistics for workplace injuries list musculo-skeletal injuries as the cause of more than half of lost-workday cases—and half of those cases involve back injuries.

Rule Number 2: Don't Up the Odds: Follow Your Procedures.

The systems and procedures you have in place are designed to reduce to an acceptable level the likelihood of the a hazard causing an event. Don't live with more risk than has been built into the process. That means, among other things, that you must follow the hazard- and risk-management procedures exactly as they have been written.

Doing that sounds simple, but hazard- and risk-management systems can often take on the appearance of a paper chase: forms to be filled out, checklists to be completed, jobs and people to be monitored. While that part of the process is necessary and adds value, the most critical parts of the process, the ones that ultimately determine the results of the effort put into managing hazards, happen out on the job. Those who do the work and carry out the procedures determine the level of success or failure in hazard and risk management.

The odds of failing to fully and completely carry out the requirements is **Execution Risk**. Many of the great failures that have made headlines suggest that Execution Risk is significant, and not limited to the people with hands on the tools. Management and managers fall victim to this risk, and often increase Execution Risk by failing to audit performance, ignoring noncompliance, not enforcing the rules, and even by signing waivers allowing *their* procedures to be bypassed.

Execution Risk is also created when people fail to fully and faithfully discharge their responsibility to independently verify the work of others. The use of crosschecks—separate and independent verifications—is a fundamental part of every operation's safety procedures. Cross-checks serve as a system of checks and balances. Most people take them for granted; some even see the process as getting in the way of getting the job done. In a sense they're not wrong. Independent verification does take time and effort, and it's tempting for people to rely on the work of others. But the logic behind independent verification is powerful. Human beings are far from infallible.

Error rates have been studied for years, and the findings are that even under the best of circumstances people routinely make errors. Our error rate rises with complexity, and under stress...and boredom. So if the procedures cover something important—like isolation—a double check by a second person is routinely required by procedure.

Statistically, the probability of two people separately making the same error (such as leaving a valve open) is the product of their individual error rates. If there's a one-in-10 chance of a person making a mistake, the odds of two people making the same mistake at the same time on the same thing are 10 times 10, one in 100. The error rate goes from 10% to 1%. It's the Law of Independent Probability.

But that's only true if both people act independently. If the second simply uses the work of the first, there is no reduction in the error rate. Asking, "Did you close the valve?" is no substitute for independently checking to see if the valve is closed.

Always "trust but verify."

The value of a second set of eyes looking at hazards out on the job follows the same logic. If you get people to first separately evaluate the potential hazards, it's likely that they won't perceive exactly the same ones. Get them to compare answers, and then it's likely that together they'll see more than either saw independently. That's *synergy*: an advantageous combination of elements.

Using two sets of eyes provides the value of the cross-check and synergy. It's a great way to increase the likelihood of recognizing what can hurt someone and coming up with a better assessment of risk.

Rule Number 3: Make Everyday Hazards—Likely the Higher Risks—a Clear and Present Danger

The logic of Dr. Sandman can be used to change the risk perception for everyday hazards. Our emotions play a big role in how we perceive risk. If hazards can be made to seem more like what people fear—graphic, dreaded, catastrophic, and controlled by someone else—they're more likely to be taken seriously. So, for example, "Treat every gun like it's loaded."

There's nothing wrong with making the case about the hazard and its probability. People know hazards—at least once they've been trained. Give anyone in your organization a written test to recognize hazards, and in all likelihood he or she will get all the answers right.

But you're probably going to have more success in getting them to recognize hazard and risk by tapping into the emotional case. Make the hazard appear more memorable and graphic. A picture is worth a thousand words: "The deck of an aircraft carrier is an accident waiting to happen." Underscore the notion that the hazard is under the control of the person: "When you don't verify the operator's preparation of equipment for maintenance before starting work, you're letting him play roulette with your life."

Rule Number 4: Reducing Risk Requires That People Be Willing to Say, "Stop."

Sooner or later, the hazard- and risk-management systems will come face–to-face with a job that doesn't meet the standards, or people who aren't following the requirements or are taking what is considered an unacceptable risk. That's the Moment of High Influence, when management's real commitment to safety shows—and when followers are paying very close attention to their leaders.

What happens then? If you are the leader—or a follower who has stopped the job because the precautions weren't sufficient or the people on the job weren't meeting the hazard-control requirements—this is the time when your actions should match your words. It's time to do what needs to be done to deal with the hazard and reduce the risk.

There is always a way to reduce risk and to perform any job safely. But that doesn't always mean the way a job is being currently done is safe. When it isn't, the first requirement is always for someone to say, "Stop."

There is a Spanish proverb, "Of all courses, the safest is always to doubt." When it comes to recognizing hazards and managing risk, a little doubt can go a long way."

CHAPTER 8

BEHAVIOR, CONSEQUENCES— AND ATTITUDE!

The actions of men are the best interpreters of their thoughts.

—*John Locke*

Chapter 6 examined the challenge of compliance: why don't followers follow all the rules all the time? When compliance is looked at as an organization behavior or system problem, the answer is found in four categories of failure: lack of knowledge and understanding, inability to remember, failure to recognize the situation calls for following the rule, and, finally, choosing not to comply. A leader who understands these four categories can better manage and improve compliance. For example, a well-executed work permit will identify the conditions that initiate specific requirements like those for entering a confined space. A Stump Speech for compliance can successfully "sell" compliance and reduce the need to enforce the rules. That's a good thing for leaders—and followers.

But nothing will ever completely eliminate the need for a leader to deal with unsafe behavior at the level of the individual.

As to how often a follower would be found working hard—but not working safely, the answer depends on where in the world you are, what work is being done, and who's doing the task. On the deck of an aircraft carrier during flight operations or when a bomb squad is examining a suspicious-looking package making a ticking sound, the odds of the safety rules being broken are small. On the other hand, watch a tire being changed on the side of the road or your neighbor mowing their yard, the probability increases dramatically. As for your followers, your answer probably falls somewhere between these two extremes.

When you come across a situation of non-compliance, what do you do?

You correct it, of course. No matter how comprehensive and well-written the safety policies and procedures are, behavior at the point of execution determines who goes home safe. When behavior doesn't meet requirements—or the work isn't being

Alive and Well at the End of the Day: The Supervisor's Guide to Managing Safety in Operations, Second Edition. Paul D. Balmert.

performed as safely as it should be—it's up to the leader to take action to change behavior. Sending people home safe means correcting unsafe behavior before it leads to an injury—not after.

As to the process to correct behavior when you see it, explaining the best practice to do so is the primary objective of this chapter. But there's more to managing behavior than simply correcting unsafe behavior. What leaders seldom realize is when they come upon a follower who is working safely, they face exactly the same decision as when the follower is not: say something or not?

If you're thinking "Why do I need to say something to someone who's doing what they're supposed to?" you're missing out on a very powerful tool of leadership in your toolbox of leadership practices: positive feedback for safe behavior. This leadership practice is so important that it deserves equal time and treatment.

What about the perceived root cause of behavior: attitude? No chapter about managing safety behavior—good and bad—would be complete without a discussion of attitude. Starting with what attitude is, and ending with what to do about the challenge of winning over hearts and minds to the cause of safety.

The best answer to that question will probably not be what you think it should be. Read on!

YOUR FIRST DECISION

The process of correcting behavior starts by deciding to deal with it. Doing so represents a decision: not every leader chooses to confront every unsafe behavior when they see it.

Overlooking the unsafe behavior of others happens frequently enough that there must be a rational explanation for it. It comes as no great surprise that people seldom appreciate being corrected. An effort to correct behavior is more often met with defensiveness—"I've seen you do exactly the same thing"—than gratitude— "Thank you for caring enough about my safety to say something." Does that always happen? No. Has it happened enough to make leaders apprehensive? Absolutely.

That's just the first entry on what is a long list of reasons *not* to intervene. It's a list worth further examination. You might be met with superior knowledge about the work being done: "You're not an electrician." Denial: "We've done it this way for years." Comparison of you to others: "Nobody else enforces that rule." Jurisdiction: "I don't work for you." Sometimes there is a compelling business reason: "If I followed the rules the way they are written, I wouldn't be finished until tomorrow. You need this done now." Severity: "This isn't that big a deal." Zero risk: "I won't get hurt." Reasons combine: people who take risk often do so in the mistaken belief they are doing the company a favor. Sometimes the leader thinks they are, too.

Then there's what might be the biggest reason: saying something feels like a confrontation. Is it?

Before emphatically replying that saying something is not a confrontation, you need to understand the meaning of the word. *Confront* is one of those words in the English language that has taken on a lot of baggage. In many circles confrontation has come to mean "getting in someone's face" … with a little "attitude" thrown in for good measure.

Saying something should not be that, because it is not that. Look up the word *confront* in a dictionary, the first thing you will see are its Latin root words, which translate into "facing against." To *confront* simply means "to come face-to-face." No more—and no less.

As a leader, should you come face-to-face with anyone who is working in a way that might send somebody home hurt? Of course you should. That's what any right-thinking, rational leader should say. Interestingly, so do followers. Ask them, "If someone saw you working unsafely, would you want them to talk to you about it?" the vast majority will say, "Yes."

It all makes perfect sense. Everyone wants to go home alive and well at the end of the day. But what those rational people—leaders *and* followers—all too often *do* in situations like these is something entirely different: they don't say or do anything. It sounds irrational, but it's really just an example of a paradox, one of those strikingly absurd, but true, contradictions in life.

Even if nobody will admit it, everybody knows all about the reasons not to say something. Recognizing that, organizations have designed around the problem, which explains the popularity of peer-to-peer behavioral observation programs. Those programs legitimize the intervention: "I'm filling out my behavioral observation card. Do you mind if I observe you?" Then comes, "I see you aren't following the rule that requires you to be tied off." That makes the intervention impersonal: "I'm just doing what the observation program requires. There's no intent to be critical of you." Even that doesn't guarantee success, because the process of being observed is usually optional. The follower responds: "I don't want you to observe me. Go find someone else to hassle."

That brings up a very important distinction between behavior observation by a peer and the intervention by the supervisor. Participation in the observation process may be voluntary; correction of unsafe behavior by a supervisor is not. Nor should it be!

Let's settle the matter once and for all. There is a long list of reasons not to intervene, and it's best not to deny they exist. But as a leader, you need to understand the reasons you must say something.

Let's start with the one reason to intervene that trumps every reason to look the other way. That's the Case for Safety: nothing is more important than going home safe at the end of the day. When it comes to safety, business takes a backseat. So do personal relationships. Looked at that way, when you say something, you're doing your follower a huge favor, whether or not they recognize it at the time.

But if that's not enough, consider the following reasons in addition to the Case for Safety.

- If you are the front-line supervisor and the follower reports to you, correcting behavior and coaching up followers is one of your job duties. Your supervisor expects you to do your job.

- You do not want to be seen by your followers as condoning non-compliance. Not saying something can be interpreted as "That's ok in my book."

- Assuming the follower knows they aren't working safely and they see you, you have been handed Moment of High Influence. And not an insignificant one, as this is a Moment created by the follower. Seize the Moment!

- Finally, you might correct the unsafe behavior, or the follower might self-correct their behavior upon seeing you. In either case, you want to say something so that their behavior is changed, not just in this moment, but in the future.

Whenever you see someone not following the rules or taking what you deem to be unacceptable risk, there is always a first decision to be made: to say something or not. Coming face-to-face is always the right decision. Choosing not to say something is also a decision. It's just the *wrong* decision.

Now that you understand why you must say something, the next question is exactly what to say in the moment. Consider that an intervention—coming between the follower and the hazard to correct behavior—and what you say is your intervention strategy.

You want to have a pre-determined intervention strategy as part of your leadership practice so you're prepared to deal effectively with a follower who's not working as safely as they should.

YOUR INTERVENTION STRATEGY

Correcting behavior is the discussion that takes place on the job, not back in the office. What may take place in the office is *corrective action:* counseling, oral or written warnings, and time off—even termination for cause. A serious violation of a life critical safety rule can lead to corrective action, and so can repeated violations of any safety rule. No matter what decisions might later be reached about the need for corrective action, unsafe behavior needs to be corrected on the spot. That is *correcting behavior.*

When you come upon someone who is working hard—but not safely—you usually have just seconds to determine how to handle what is an important and often sensitive conversation. What if you don't have a pre-determined intervention strategy, instead choosing to "think on your feet" on the theory that no two situations will be alike?

If you're upset, you won't do your best thinking. You can easily get sidetracked by what your follower has to say. You might go down the path of discussing something that doesn't matter. You might get into an argument. You won't know what to say that will succeed in changing your follower's behavior.

On the other hand, if you know what to say to do that, you have an intervention strategy: what to say; the order to say them, and why to say these things.

So, what is the best practice to correct behavior?

Every parent—or child, for that matter—is well schooled on the subject of correcting behavior. We've grown up with it; raised our kids practicing it. You couldn't have played any kind of competitive sport without being on the receiving end of behavior correction from your coach. We've all witnessed plenty of models for correcting behavior, some better than others. In most respects, correcting safety behavior isn't any different. Yes, you're dealing with adults—but we're all just grown-up children, at work, not out on the playing field. The basic principles that apply universally fit unsafe behavior—with one very significant difference.

The first principle is obvious: when correcting behavior you should always *focus* on behavior. Behavior is action, what a follower is doing, or not doing. Safety behavior is what you see: not wearing a hard hat, leaving earplugs dangling on the shoulders, working with an unsigned permit, not wearing a respirator, straddling a hose, leaning over a drum, lifting improperly. Behavior is found in the facts of the situation; facts are what can be proved by evidence. When correcting behavior, you always want to make the facts known, and clear. Doing that moves the focus from the person to the facts, and lessens the tendency to argue back: it's difficult to mount a good challenge against facts.

In the heat of the moment it's easy to begin judging behavior: it's willful, irresponsible, unprofessional, careless, or just plain stupid. Leave those adjectives out of the conversation; stick to the facts. There is no upside to throwing those kinds of judgments into the conversation: they only inflame passions.

The second principle: compare the observed behavior to what was required or expected. Doing that establishes the performance gap: what was observed and what was expected. The objective of correcting behavior is to close that gap.

Safety policies and procedures establish requirements: hard hats are to be worn in the shop; a permit must be completed before maintenance work can be started. An injury is to be reported to the supervisor immediately. Requirements go well beyond the written safety rules. They may also come in the form of training about safe practices or proper workmanship: don't stand in the line of fire; use proper lifting technique; read the label before opening. There are plenty of ways to get hurt that don't involve breaking the safety rules. Consider them expectations: hold on to the handrail, look both ways before crossing the street. Not meeting expectations also establishes a gap, making that fair game for correction.

Describing behavior and stating the requirements are two key elements in an intervention strategy. Together they establish the performance gap and both are facts: things that can be proven. Nobody can argue with the facts.

The third principle: ask why. "Why aren't you following this rule?"

Many leaders—and many behavioral observation processes—skip this step. The justification is that people know the rules, and the leader doesn't want to hear any excuses as to why those rules aren't being followed. That might be so, but there is so much power found in that simple question: asking why is the single best thing a leader can do when intervening. Consider the many benefits of asking the question, why.

A question changes the nature of the intervention, from a monologue to a dialog. Asking why doesn't just invite participation, it creates an expectation there will be involvement. The follower is not allowed to passively receive a lecture from the leader.

There must be a reason why the follower wasn't working safely. If you are the leader responsible for the person's safety, don't you want to know the reason for the behavior? That might make changing it easier.

There could be a very understandable (and possibly even acceptable) reason for the behavior you observed. "The reason I'm not wearing my earplugs is that we've run out and they're on back order." Knowing that doesn't change the fact that the rules weren't being followed, but it does provide some very useful intelligence about what needs to be done to change behavior. It also might create work for the leader.

Be forewarned: when you ask for the reasons for the behavior, what you hear won't necessarily be the truth. "I forgot and when I saw you, I suddenly remembered." That's nothing more than the adult version of the oldest excuse in the book: "Teacher, a dog ate my homework on the way to school." Getting excuses instead of the truth can be very frustrating, but even when you're getting a lame excuse, you've succeeded in making it a two-way conversation. When you ask a question, someone can't just nod and pretend to listen.

More importantly, asking *why* is a very effective tool of influence as the question causes the follower to examine the reasons for their behavior. (You'll learn more about the power of questions in the next chapter.) No matter the answer, the leader has forced the follower to examine their logic and decision-making process. That is a key step in changing future behavior.

As to what a follower might say in response to the question, it can only be one of two things: the truth or not. Getting the truth, no matter how ugly that might be, is always beneficial to the cause of safety. When it's not the truth, remember that asking adults to own up to their failure is a lot to ask.

The fourth principle: explain consequences. Consequences drive behavior. Not following the safety rules or working unsafely puts two potential negative consequences into play: the possibility of getting hurt, and the possibility of getting into trouble.

Either of these two potential consequences can change behavior, but one is far more impactful and long lasting than the other. Getting hurt—or hurting someone else—is always the greater consequence, the one people should fear more. That's the Case for Safety. Suffering a life altering injury produces consequences someone must live with for the rest of their life. The administrative consequences of corrective action pale by comparison. It might amount to little more than getting a lecture from the boss. Everyone gets over that. At its worst, it could mean loss of a job. There will always be other employment opportunities.

When neither of these two consequences is believed to be in play—the person performing the task is absolutely certain they won't get hurt, and very sure they won't get caught and get in trouble, an alternative set of consequences will drive behavior, and will inevitably drive it in the direction of noncompliance.

Every safety rule represents some kind of encumbrance to getting the work done. That's why the rule was written: to make people do things differently than they otherwise would if left to their own devices. Consider all the extra effort involved in following the rules: it takes more time, slows the job down, and requires more effort—mental in the case of procedures like the job safety analysis, physical in the case of proper lifting. If nothing else, on a hot afternoon, wearing all the required PPE is just plain uncomfortable. If there isn't any perceived benefit to putting in all that effort to comply, why would any rational person suffer all the negative consequences of *following* the rules?

The solution to this inherent problem with the safety requirements is to explain the benefits of compliance. The consequence of getting hurt—or of going home safe—trumps all the consequences of having to follow the rules. Every safety rule and expectation is there for a reason, and the principal beneficiary of working safely is the person doing the work. After all, those are *their* fingers and *their* eyes: were something to go wrong, *they* are the one who would go to the emergency room for treatment. The boss just goes along for the ride. Don't expect someone to accept that as an act of faith, or assume they understand it. Part of correcting and influencing behavior is to explain the benefits from safe behavior by explaining how someone was hurt in a similar situation, or was saved from injury by following the proper procedure. It should be easy for you to come up with examples based on experience.

It's tempting to threaten corrective action: "If you get caught again without your PPE…" The problem with corrective action as the primary deterrent is that the threat exists only to the extent management is capable of enforcing the policy. By comparison, the risk—probability—of an injury doesn't depend on the presence of management and its commitment to enforcing the rules.

Connecting up these four principles, you have the elements of an effective strategy for correcting behavior:

- describe the behavior
- state the requirements and expectations,
- ask why
- explain consequences.

There's only one thing missing from an intervention strategy: how to start the conversation.

That's often the toughest part of confronting behavior. Most of us don't like being criticized, and many of us aren't any too anxious to criticize someone else. We all know we're far from perfect. In our attempt to be courteous and respectful, and not to put someone on the defensive, the conversation may start on a positive note. It might be small talk: "How's the family?" "What's going on with the job?" A compliment: "Hey, the place looks great. Thanks for all the hard work. Now about not wearing your safety glasses…"

It's an approach that seems to make confrontation easier on us, and reflects a popular approach to giving feedback: positive first, negative second. Then, finish the conversation on a positive note.

The theory behind this positive/negative/positive approach is that by starting with the positives, it establishes and preserves a good relationship with the follower whose behavior is being corrected and makes that person less defensive when you get around to delivering the bad news. It seems to make sense, until you consider what's really going on when the leader follows that model.

The first problem with the theory is that it has been practiced for so long followers have figured it out. The leader is hard at work giving the compliment and saying kind things, while the follower waits for the inevitable criticism. It always follows: that is the model.

It's doubtful the followers see the compliment as being real and sincere. That does not enhance a leader's credibility.

If it were simply a case of a few wasted words, taking that approach wouldn't matter all that much. But there is a more critical failing: while the behavior involves safety, the compliment usually involves some other aspect of the job. . .getting the product out, getting the work completed, satisfying the customer, cleaning up the yard. Telling someone who is taking a risk that the leader appreciates the hard work they did before discussing their lack of safety compliance could well give the impression the leader thinks taking care of business is more important than working safely.

Is that what the leader thinks? Likely not. Is that what the leader wants to convey? Hardly. Is that the message heard by the follower? Very possibly.

What the follower needs to hear is that despite what else may have gone right on the job, if they were to have been hurt none of those good things matters. That's the Case for Safety.

The best way to correct behavior is to get right to the matter at hand. What is needed at the beginning is called an icebreaker: a way to get what might be a sensitive conversation started off on the right foot. In every situation, there's always a simple and obvious way to start: telling the follower why the leader is there. There's always a reason, it can be said in a sentence, and it doesn't matter. It just does no harm.

In the case of the shop supervisor's visit to the cleanup job in the yard, the icebreaker goes like this: "I came by to see how the cleanup job was going and how safety looked out here in the yard."

THE SORRY MODEL

Putting these five principles of correcting behavior together in a logical sequence creates a simple five-step intervention strategy:

1. Start the conversation by **stating** why you're there.
2. Describe the specific behavior you **observed** is the problem.
3. State the **requirements** for performing the work safely.
4. Ask the **reasons** for the behavior.
5. Explain what can happen to **you** if **you** don't work safely.

That might seem like a lot to say, but first three steps simply involve stating the facts: why you are there, exactly what you observed, and what you should have seen. That requires no more than three sentences. Asking why is easy; dealing with the answer may not be nearly so simple. As to effectively describing the potential consequences—what could happen—that requires you to draw on your experience, what you've seen happen to others, and perhaps even to you.

Those five steps might seem like a lot to remember in the heat of the battle, but if you look closely at the words in bold—state, observe, requirement, reasons, you—there's an acronym to remember them: SORRY. The SORRY model is the best practice for an intervention strategy to correct unsafe behavior.

Had this model been followed, here is how the conversation between the supervisor and the two crew members assigned to clean up the yard—who put on their personal protective equipment only when they saw the boss headed their way—might have gone.

2:05 p.m., Back of the Yard

As their supervisor approached, Ron and Charlie were intently focused on sorting parts, barely looking up in the hope nothing would be said about that little problem of noncompliance. The yard was clean, and they were now wearing the required PPE.

That hope was dashed when their supervisor started the conversation.

"I came by to see how the cleanup job was going, and while I was at it, to see how safety performance looked out here. When I first saw you, neither of you were wearing your hard hat, safety glasses or gloves. Now they're all on. You know the rules: personal protective equipment is required anytime you're out of the office and working.

"So, what's the story? How come you're only putting your PPE on when you see me coming?"

There was a long pause. The leader was determined to hear their reply, and waited, waited, and waited.

Finally, Charlie broke the silence. "Well, you know, we'd been wearing them all day. When we came off our break, we just plain forgot. When we saw you, we remembered and put them on."

The supervisor smiled. Back in the day, they tried that line on their supervisor a time or two; it didn't fly. Lesson learned. The real story wasn't hard to figure: in all likelihood, Ron and Charlie never expected their leader to show up on a clean-up job, at least not at 2 in the afternoon. The supervisor filed that lesson away for future reference.

A little bit of re-education about the hazards of the job was in order. "Look, I know on a hot day like this it might be tempting to take a shortcut on the PPE requirements, particularly when you think this is 'just a cleanup job.' Back when I had your job, one of the worst accidents I ever saw had happened to one of my peers in the machine shop in a situation just about like this. Let me tell you what happened…"

Moments later, the leader had the two won over. "Sure, if you get caught again, you might be in some hot water. But that would pale in comparison to what happened to one of my co-workers. You don't ever want something like that to happen to you."

Behavior corrected.

THE POWER OF POSITIVE REINFORCEMENT

Considering the caliber of your followers—good people—and the level of safety performance they're achieving—good, as well—come time to perform Managing By Walking Around, you're likely to find good, safe behavior. Should you say something in that situation?

Often the answer to that question is, "No."

Whether or not you do, the first thing to understand is that you're facing exactly the same choice—a decision—as you do when one of your followers is not working safely: say something or not. For the reasons described earlier, you must say something to the follower who is not working safely. You aren't required to say anything to followers who are, but you need to recognize you have that option.

Many leaders choose to say nothing. As to their logic, it goes as follows. Working safely is what is expected. A leader shouldn't have to praise someone for meeting the basic requirements of the job. When people work safely, they are rewarded: they don't get hurt. Safety is factored into their bonus compensation, so there is even a financial reward for doing the right thing. Praise should be reserved for going above and beyond the norm. Start praising people for what's expected, where do you draw the line? Praising people takes time. There are more pressing matters, dealing with problems. So, "If it ain't broke, don't fix it."

The more formal name for that approach to supervision is *managing by exception*. It's a well-practiced model for allocating the scarce time of the leader: pay attention to the important problems, those representing the "critical few" that spell the difference between success and failure. The things that are going well don't require that most critical of resources, the time and attention of the leader. In a sense, Managing by Walking Around is based on that premise: determining the most likely reasons for people to get hurt, and focusing time and attention on the situations in which those reasons would most likely be in play.

But acting on managing by exception as a primary leadership practice can produce negative side effects, one of which is that when the leader shows up on the scene it's never a good thing—because there must be a problem. When a leader argues that the process of correcting behavior must begin by giving a compliment so that followers will listen and not become defensive, it's a clear indicator the leader isn't doing enough in the way of recognizing good behavior. If it were otherwise, the leader wouldn't feel compelled to "lead with a positive."

Observing the issues created by managing by exception, a consultant named Ken Blanchard, known as the One Minute Manager, coined the phrase, "Catch them doing something right." Blanchard's observation was that positive feedback was just as effective in promoting good behavior as correcting wrong behavior. And, as he wisely noted, there is no downside to giving a follower a compliment.

Some of the best leaders on the planet had learned that lesson long before the One Minute Manager arrived on the scene. One was a basketball coach, Larry Brown, who was recognized for his tremendous success both as a basketball coach and as a long-standing practitioner of what he called "positive coaching." Here's what the Coach had to say about his approach to giving his players positive feedback:

> When I was coaching in college, there were students doing a thesis. They came to our practices and recorded my responses, both positive and negative. My ratio was 4 or 5 (positive) to 1 (negative), which was high compared to other coaches. I was surprised because I feel like I am constantly on the guys. I want my players challenged, and I'm not afraid to be tough. But you want to bring out their best. My biggest challenge is letting the guys know the difference between coaching and criticism. You have to make them understand that you are trying to make them better.

Here's a successful and demanding coach who's won championships at the college and professional level giving his players far more positive than negative comments. Leading followers is a lot like coaching players, and positive feedback makes an impact.

When giving followers positive reinforcement for safe behavior, there's a sense that this is done for the benefit of the follower. That is certainly true: the follower is the biggest beneficiary of going home alive and well at the end of the day. That is the Case for Safety. What are routinely overlooked are the benefits that accrue to the leader by giving a follower a compliment for working safely:

- It creates a Moment of High Influence.
- The follower is more inclined to work safely.
- When a follower tells peers "I got a compliment from the boss" it motivates peers to earn similar recognition.
- Followers know that their leader is paying attention, and what the leader is paying attention to.
- Relationships improve.
- The trust and respect of the leader grows.
- Followers more readily accept coaching and behavior correction.
- The leader feels better.
- Managing safety performance becomes easier.

This is an impressive list of benefits—to the leader! You don't need or want to give positive feedback to every follower every time you see them working safely. You want to pick your spots: give the recognition where and when and to whom you believe will serve your purpose.

GIVING POSITIVE FEEDBACK

If you are now persuaded that giving positive feedback is a good management practice, and you want to give a follower a compliment for working safely, what exactly do you say? You could simply note, "Well done." "Looking good." "Thanks for working safely." There's always the thumbs-up sign. Doing any of those things would be a step in the right direction, but they fall short of gaining the full benefit of positive reinforcement.

If you want your words to influence future behavior, there are three principles for giving positive feedback that will make a difference in future behavior:

1. Be specific about the behavior.

The best sports coaches know that principle: when they're coaching athletes in practice, they don't just say, "Nice play." They describe the player's specific behavior. The golf coach tells the golfer, "Way to keep your head down well beyond impact." It isn't necessary, or even useful, to describe every facet of the correct behavior in play. The message of positive reinforcement gets across better when the focus is on the specific desired behavior. Pick your target: if improving compliance with the hearing protection requirement is the focus, the observation that earplugs are being properly worn should be specifically described.

2. Be sincere.

It's easy to see right through a compliment that isn't genuine; follow a formula to give positive feedback and you run the risk of seeming false. The way to avoid that is to make sure the positive reinforcement is specific to the individual. There is always some specific aspect of positive reinforcement to be found: in the task, the time of day, the circumstances surrounding the task, the person's track record. In the case of the cleanup duty at the shop, had the rules been complied with, that specific aspect could have been presented this way: "I know it's a hot day and you're just cleaning up the yard. But following the PPE rules out here sets a great example for everyone else inside the shop." Reinforcement focuses on the individual; keeping that in mind will always make a compliment sincere.

3. Sell the positive consequences of working safely.

That might seem to go without saying, but after all the effort to improve safety performance during the last hundred years, there are days when the followers doing the work think they're doing *their leader* a favor by working safely. Normally, the sales pitch for working safely is reserved for the situations when they aren't. Why not sell when they're already buying? That guarantees no resistance. What's in it for someone to comply? "Even though it's just a cleanup job, and you're working outside, we've had more than our share of hand and eye injuries outside. Wearing your PPE will make sure you go home safe in case something unexpected happens."

The best time to give positive feedback is when you see the behavior. If you're looking for an icebreaker, there's an easy way to start the conversation: telling them why you happen to be there. That's exactly the same approach used when correcting behavior. When reinforcing good behavior, starting with that reason avoids confusing the message with small talk or discussion about the state of the work. Save that for after the compliment has been given. Finally, it never hurts to remind people of what the requirements are—even when it's clear that they know.

Putting the principles in a logical sequence provides a simple, five-step strategy to provide positive reinforcement of safe behavior that is nearly identical to the intervention strategy for correcting behavior:

- Begin by **stating** the reason why you're there.
- Describe the specific behavior you **observed**.
- Remind them of the **requirements** for performing the work safely.
- **Reinforce** the behavior in a way that makes the compliment genuine.
- Sell the benefits—what's in it for **you** to work safely.

As to what's different, when a follower is not working safely, always ask why. When a follower is working safely, it's not a good practice to ask that question: you might be disappointed by their reason for compliance. So, don't ask. Instead, reinforce the good behavior.

In practical terms, reinforce means making the compliment come across as real and genuine: the feedback is about the follower, not the leader. Reinforcement is about the person in the situation—not about what can happen by not working safely. Those are consequences. As to how best to reinforce good behavior, simply consider factors such as the person, the situation, the requirement, and their history with working safely.

Had the shop supervisor found good compliance on the part of the two crew members assigned to cleanup duty, the conversation might have gone something like this:

2:05 p.m., Take Two

The supervisor approached Ron and Charlie, who were intently focused on sorting parts. They looked up, surprised to see the leader out in the yard at that time of day. The leader was impressed with the progress they had made, evidence they were working very hard. More importantly, both were working safely. The supervisor started the conversation.

"I just came by to see how the clean-up job was going, and while I was at it, to see how safety performance looked out here. I can see both of you are wearing your hard hats, safety glasses and gloves. That's exactly what's required for this job.

I know it's a hot day and you're just cleaning up the yard. Following the PPE rules out here sets a great example for everyone else inside the shop. I'm really glad to see someone with your experience, Charlie, following the rules and setting a good example. Ron, good to see you not only knowing the rules, but also the habit of following them. It's a great way to start your career.

Even though it's just a cleanup job, and you're working outside, over the years we've had more than our share of hand and eye injuries in the yard. Plus, it's easy to get complacent on what might seem like a routine task like cleanup. But injuries are more likely to happen we let our guard down. Wearing your PPE will make sure you go home safe in case something unexpected happens."

Five steps might seem like a lot of effort to go through to give a compliment. But the whole conversation will take less than a minute, and the key elements of giving a compliment are well executed: behavior observed, requirements that apply, reinforcement to make it genuine, and positive consequences described.

MANAGING ATTITUDE

When you see unsafe behavior, as a leader you're obligated to correct it. When you see someone working safely, it's a terrific leadership practice to reinforce it. What about managing attitude?

In the opinion of the overwhelming majority of leaders the world over, in managing safety performance attitude is every bit as important as behavior. Many leaders will passionately make the case that attitude is even *more* important than behavior! As you move up the chain of command in the direction of the executive suite, that conviction only becomes stronger. If nothing else, it explains the interest in values, the focus on winning over hearts and minds, the emphasis on creating buy-in to become injury free, the significant investment in attitude surveys, and the posters on the walls proclaiming, "The A in safety stands for attitude!"

If you're running with the herd on the matter of attitude—it's a herd usually in stampede mode—now is the perfect moment to stop and ask yourself a few questions about the tough challenge of managing the attitudes of followers.

- Why is attitude that important—to you as the leader and to safety performance?
- How do you manage—and, more to the point, *change*—the attitude of your followers?
- How do you know what a follower's attitude really is?
- When you say "attitude" what do you actually mean?

Any leader serious about taking on the challenge of managing the attitudes of their followers needs to not just reflect on these four questions, but have the knowledge and understanding as to the correct answers to each of these questions.

Sadly, most do not. Were they to understand what is required to successfully undertake managing the collective attitudes of their followers, many leaders would think differently about the solution.

And the problem.

WHAT EXACTLY IS "ATTITUDE"?

Every college freshman who's taken an introductory course in behavioral psychology will be asked to define attitude on their first-six-weeks test. According to the Swiss psychiatrist Carl Jung, an attitude is "the readiness of the psyche to act or react in a certain way." That answer will earn a passing grade on an exam, but "readiness of the psyche" is not the least bit helpful to a supervisor or manager trying to get followers to quit complaining about the rules and start following them. Attitude can be defined as reflecting the sum of one's life experience that leads to some predisposition. It might be considered as the combination of values, beliefs, principles, and assumptions. That doesn't help matters. Out on the shop floor, the practical definition of attitude usually goes, "What someone is thinking."

You can take your pick of these, or any number of other definitions of attitude. This is not an academic book, nor is your interest in the subject academic. If you are of the view that attitude is that important, let's not overcomplicate matters, and define attitude as simply "what someone is thinking,"

The important point made by this practical definition (or any textbook definition) is that attitude, unlike behavior, is not something that can be directly observed and measured. Attitude is bound up in the five and a half inches between the ears.

That definition begs a question: how do you as a leader know what a follower is thinking? That is not an inconsequential question. Quite the opposite: in order to manage something, you have to be able to measure it.

How do you measure what's going on in the minds of your followers?

ATTITUDE SURVEYS

We are but two questions into the subject of attitude, starting at the bottom and working up towards the top. What is the definition of attitude? What someone is thinking. As a leader, how do you know with certainty what your followers are thinking? That you can only guess.

You can begin to see the problem you face when attempting to manage attitude. But you might also be of the view that you can find out about attitudes by listening to what your followers have to say, and by asking them in the form of an attitude survey.

While measuring what people are thinking is done all the time in the form of what are commonly referred to as attitude surveys, the underpinnings of attitude surveys are very complex. Safety surveys are more than complex; they can be downright misleading.

Here is an example that makes that point clear. Consider these two questions that typically show up on safety surveys, paired with a range of responses running from strongly agree to strongly disagree. It's a good bet you've been asked them yourself.

1. I would say something to someone I saw taking unacceptable risk.

2. My supervisor always puts safety ahead of production, cost, quality, and schedule.

That there is a "right" answer to each of these two questions is obvious: everyone should always intervene, and every supervisor should always put taking care of safety ahead of taking care of business. The question is: do the follower's answers actually tell you what their practices of intervention and the priority given to safety by their leaders are—or something else?

If you wanted to determine with a degree of certainty the actual practices in an operation, that would require observation. Do that, and you would know the facts. That being the case, what's the point served by asking followers what they *think* is the practice? Their perception may or may not accurately reflect practice.

As to how they might reply, put yourself in the mind of two different followers asked those two questions in a survey.

Follower A is a conscientious member of the outfit who wants to do the right thing, and isn't out to make anyone look bad. The follower might very well answer both questions correctly—as defined by their leaders. The correct answers are obvious, aren't they? For the first question, they know intervening when someone else is taking unacceptable risk is the right thing to do; for the second question, the follower sees their supervisor as a good leader.

That follower tells you, "What they think they think." But when faced with a decision about intervention, they still might not intervene. Good intentions notwithstanding, in many places intervention is not the norm, for all the reasons described above.

As to the second answer, does it reflect what this follower's supervisor actually does in practice—putting safety first? Or is it simply a perception, formed on the basis of the age-old "halo effect": "I like my boss, and I'm sure he or she will always do the right thing."

This should expose the truth about attitude surveys: results provide very little information that can safely be relied on.

If you still need to be convinced, consider Follower B, who knows from experience a survey is a way to manipulate the system. What might they be thinking—and answering?

The possibilities are limitless:

- Still mad about a decade-old safety problem, the follower decides answering the questions "wrong" will send a message to upper management—and get back at the supervisor.

- The follower thinks making his boss look bad on a survey might lead to getting a different supervisor...one less inclined to make the follower put on their safety glasses and hard hat.

- The follower has learned that making the department look good on surveys keeps the corporate people off their backs.

- The follower knows that a high score on the safety survey counts toward the next safety bonus.

Now you see the point very clearly: how people respond on a survey may reflect exactly what they think, and it may not. How does someone who reads the survey findings know the difference? The experts in the business will tell you they can interpret the findings and tell you what the truth is. Part of that truth is that conducting surveys is a very profitable business—for the people who ask the questions.

But you're not in that business. As a leader you ask followers what they think—and listen to their answers. As long as followers tell you what they're really thinking, the approach works well.

But do followers always tell you what they think, face to face? Of course not. How do you know when they are telling you what they really think? Or when they're telling you what they want you to think they think? Or if they even know what they really think?

It's enough to give a leader a headache. But if you're serious about wanting to manage attitude, you have to think about these things.

Malcolm Gladwell nicely summed up the challenge in his best-selling book, *Blink: The Power of Thinking Without Thinking*. Figuring out what people really think sounds easy, but proves extremely difficult for a long list of reasons. Chief among them is that people don't always know what's going on inside their own heads.

That explains why opinion polls and focus groups aren't necessarily an accurate reflection of what's going on inside the heads of followers.

THE EXPERTS SPEAK

Richard Beckhard was a long-time faculty member at the Sloan School of Business at MIT and a very successful management consultant. In the 1970s, when cockpit miscommunication was identified as a significant factor in several high-profile commercial aviation accidents, Beckhard was one of those hired to do the research and solve the problem by what became known as Cockpit Resource Management.

Beckhard's genius was in his ability to reduce very complex subjects to simple, practical, understandable—and workable—terms. His research method? As he put it, "How can you see what's going on without actually being there?" Beckhard flew in the jump seat in the cockpit—that's the very small space immediately behind the captain and first officer, taking plenty of notes on plenty of commercial flights. It was a brilliant idea—coming from a guy who could play Santa Claus in the Macy's Thanksgiving Day Parade. Imagine Santa buckled into the jump seat, and you have an idea of what this must have looked like to the pilots he was observing.

A keen observer of human behavior, Beckhard didn't disagree with the premise that by changing attitude a leader would address the root cause of behavior. But he posed two questions: "How can you know what someone else is thinking?" And, if you think you do, "How do you change what someone else thinks?"

Long before Richard Beckhard, Peter Drucker described the problem of managing attitude this way: "Nothing is more difficult to define...difficult to change." He wrote that in *The Practice of Management* in 1954.

If attitude can't be accurately known in the first place, how can it successfully be changed?

For the sake of the argument, let's assume you are convinced you know exactly what your followers are thinking (even if you never really will know for sure.) You work closely with your crew; you've known everyone on it for years, there's openness among the members of the team that assures you that you know what they think—not through survey findings, but by firsthand experience. You probably do have a good idea what your followers are thinking.

If so, how much influence can you have over what they are thinking? Likely some, but how much?

Here's a very simple way to determine the answer to that question. Ask yourself how much influence someone else can have on your attitude. Your answer depends on a great number of factors...who the person is, what his or her relationship is to you, the subject in question, how strongly you feel about the subject, the techniques the person employs to convince you to change—frankly, even what kind of mood you might be in at the time. These same factors—and likely many more— are in play when you try to change a follower's attitude. The odds of a significant change in what your followers think aren't good. Why would you choose to invest your limited and valuable time in the pursuit of a goal for which you are highly likely to fail?

Long before Beckhard and Drucker, one of the great philosophers in the Age of Enlightenment wrote: "The actions of men are the best interpreters of their thoughts." That's how John Locke put it—in the 18th century.

MANAGING BEHAVIOR

So now we're back to where we began: why is attitude that important to leaders? The answer is simple: because attitude drives behavior. Get the attitude right, and behavior is sure to follow. Moreover, safe behavior will occur whether the leader is out on the floor or back in the office entering the day's production into the computer.

In theory, it's the right answer. The problem is with putting the answer into practice. Leaders who invest a great deal of their time and energy trying to manage and change attitude often have little to show for the effort. They can't know for sure what the attitudes of their followers are, and even if they did, changing how their followers think falls somewhere between difficult and impossible.

Investing time on something that probably won't work isn't a smart strategy. If you're a leader depending on attitudes to change in order to improve safety performance, you're setting yourself up for a huge disappointment. Follow that strategy, you will continue to suffer poor behavior—and poor safety performance—while you're waiting for the attitudes to change for the better. How much time do you have to see if attitudes improve, and if better behavior follows?

Not nearly enough!

By comparison, managing *behavior* is simple and straightforward. Behavior takes no genius to decipher and interpret. Behavior is found in actions: what people do. See it, and you've got it. Behavior can be measured by observation. Behavior can be directly influenced by managing consequences. If you're a busy leader, managing behavior and consequences represents a far more straightforward and productive use of your time and resources. Your job is already tough enough…why complicate life by taking on attitude?

The model that we've described for correcting and reinforcing behavior is built around behavior and consequences. Yes, attitude drives behavior—but so do consequences. Attitude notwithstanding, people will do the right thing because they will receive positive consequences: wearing earplugs will save their hearing. They will do the right thing because they don't want to suffer negative consequences: "If I get caught without my earplugs, I'll be in trouble with the boss." Positive consequences reinforce behavior; negative consequences alter future behavior. The relationship works in both directions. See Figures 8.1 and 8.2.

That behavior produces consequences, and consequences alter behavior comes as no great revelation to any leader. It's the logic underpinning performance appraisals, compensation systems, safety bonuses, and, yes, corrective action policies. In the SORRY Model, it is the "observation" and the "what can happen to you" steps: behavior can be seen and consequences can be explained and administered by the leader.

But there's an interesting connection many leaders don't appreciate: behavior can change attitude.

Just as the relationship between behavior and consequences works in both directions, the same holds true for the relationship between behavior and attitude. Behavior can determine attitude, in just the same way that attitude can drive behavior. The driving force here is called *cognitive dissonance*, the uncomfortable feeling you get when you hold conflicting beliefs. It was so named in 1957, but the phenomenon has been around just about as long as have we humans.

Long before there were behavioral psychologists, the military figured out the genius in this. Yes, the attitude of new recruits is very important to their effectiveness in combat. But what militaries the world over became world-class at doing is managing *behavior*, particularly during basic training. Their approach, honed over years (more likely centuries) of experience, is to manage the behavior of new recruits in

THE INTERRELATIONSHIPS AMONG BEHAVIOR, CONSEQUENCES AND ATTITUDE

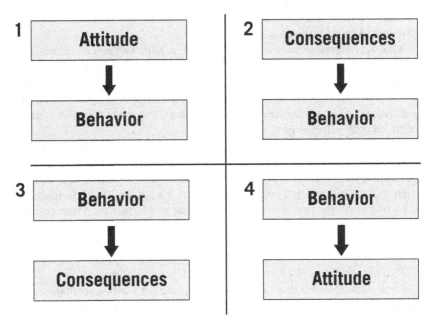

Figure 8.1   Attitude, behavior, and consequences are all intertwined.

THE ABC MODEL

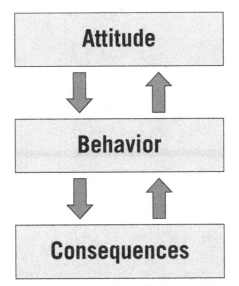

Figure 8.2   Managing consequences to drive behavior can ultimately change attitude.

excruciating detail, and to give constant feedback about that behavior. By experiencing that intensely as part of the basic training experience, the recruits wind up changing their attitudes to match their behavior, in order to get rid of that feeling of cognitive dissonance. Within a few short weeks cleanliness, orderliness, and following all the rules all the time become vitally important to that same 18-year-old who had to be reminded 15 times to clean up his or her room.

What is achieved in military basic training is nothing short of stunning. In a matter of weeks, a diverse group of inductees are given a new identity, a set of common values, and a different code of behavior. New recruits rapidly assimilate values and traditions that have existed for literally hundreds of years in their branch of the service. The model of military basic training suggests that when behavior is actively managed, attitude will change to fit the behavior.

This is not to suggest the same regimen for your operation. In basic training, early on very little feedback is positive. Leaders in industry don't have the luxury of a captive audience; followers can check out. But you can learn from the successful experience of the military and apply their wisdom: it's not the speech from the commanding officer on the first day of boot camp that makes recruits Marines, but the constant pressure from everyone in the system every waking hour that forms and influences behavior. In turn, that constant and intense pressure on behavior creates the desired attitude.

It's a model that works for the military, and it can work for you.

Finally, any discussion about attitude must be ended with a reminder of what makes the difference in working safely. It is *action*: what followers do, and how well they do things that ultimately determines whether they go home, alive and well at the end of the day.

Action is always the prize: keep your eye on the prize!

CHAPTER 9

THE POWER OF QUESTIONS

Ask questions when you don't know the answer, and sometimes when you do.

—Malcolm Forbes

Peter Drucker described leadership as "making common men do uncommon things." When it comes to safety, that uncommon thing is getting people to work far more safely than they would do if left on their own. Achieving that level of performance is hard work, requiring a huge investment of the time and effort on the part of the leader. One way to reduce the amount of leadership effort required—or to get better results for the same effort—is to look for leverage. Mechanical advantage is one of the oldest and most basic principles of physics. One management application of that principle is asking questions.

Leaders ask their followers questions all the time. "How much did we produce?" "Why is Line 6 down?" "When will it be re-started?" "Has the order been shipped?" Questions like those have two things in common: they're all questions, and they're all questions in search of information.

Most of the time, when a leader asks a follower a question, the objective is simply to get information. There's not a thing wrong with that. Leaders need information and when the follower has it, by all means, ask. But there can be an entirely different purpose for a question posed to a follower: not to gather information, but to gain influence.

Asking questions as the means to influence followers is one of the oldest leadership practices on the planet. The philosopher Socrates was the master of the technique: "I cannot teach anyone anything, I can only make him think." Socrates did that by asking questions. Not questions for information or easy-to-answer questions with right and wrong answers, but tough, thought-provoking questions—questions that forced a full examination of the matter at hand, with the potential to change people's minds. It was the stuff of genius: upwards of twenty-five hundred years later, his questions are still being debated.

Alive and Well at the End of the Day: The Supervisor's Guide to Managing Safety in Operations, Second Edition. Paul D. Balmert.

Using questions to lead and influence followers can be one of the most powerful tools a leader can employ to manage safety. Among the many benefits is that when the leader is asking, the follower is doing the hard work. As former CEO Larry Bossidy described it, "Let your questions do the heavy lifting."

Accomplishing that by asking questions requires understanding on the part of the leader as to technique: how to ask the kind of question that will cause the follower to willingly answer, and in so doing lead themselves exactly where the leader wants them to go.

As simple as "leading by asking questions" might appear, doing it well is not simple stuff.

THE POWER OF THE QUESTION

Consider what happens when you're asked a thought-provoking question. For example, "What's the biggest risk you've taken in the last 24 hours?"

Asked a question like that, you can't help but to begin thinking about the subject. You review the work you've done in the last day. Then you might expand the scope to include what you did off the job, and on the way to and from the job. You begin to evaluate, comparing all those activities based on how you define risk. In your mind, you develop a list, and then reorder the list based on what your experience tells you is truly "risky." Perhaps you start to rethink what the word *risk* really means, and maybe even how you know what the risk is for anything you do. That deliberation in the mind produces an answer: your answer could be, "Driving home at the end of the shift."

All that takes place in what seems like the blink of an eye; it pretty much is that. Rarely does anyone stop long enough to realize what just happened inside their heads: a simple question causes someone to become fully engaged with—and thinking about—a subject. In that process there's analysis, reflection, discovery, and evaluation. An opinion might be formed or changed; something might be learned or understood; the motivation created to learn or find out.

This is powerful stuff, going on in the five and a half-inch space between the ears, caused by simply asking a question. The work is done by the follower. That's the kind of leverage a good question can have.

Alternatively, someone could have just told you, "Statistically, the drive to work is the greatest risk you'll take today." You might argue the point: perhaps you rode your bike to work. But even if you were to accept the statement as factually correct, you won't own the information nearly as well as you would, were you to have reasoned your way to that conclusion.

That's the genius of a question, something that Socrates figured out a couple of millennia back. There *is* nothing new under the Sun. But doing that takes the right question, asked in the right way.

Asking the right question the right way is a very effective way to get followers engaged, thinking about what they're doing, and focused on the task at hand, learning, complying and even to communicate important information, all of which show the list of the toughest safety challenges a leader faces.

Leading by asking questions works so well you would think its use would be commonplace. But in the practice of leadership, leading by asking questions is relatively rare. Why so? The reasons aren't hard to fathom. Leading via asking questions requires a leader to be willing to:

- Listen to what the follower has to say
- Hear things he or she might not want to hear
- Convey the impression the leader doesn't have all the answers
- Invest precious time in the process of dialogue

Bottom line, leading by asking questions requires the leader to hold his or her ego in check— and have the patience to let the follower do the talking. But if you are looking for one more reason to be convinced it's worth the effort, remember every leader is also a follower—of some other leader. You can't find a leader who doesn't think they have important opinions the boss should hear—if only their leader would take the time to ask, and be willing to listen to them.

The greatest management consultant of the twentieth century, Peter Drucker, said of himself, "My greatest strength as a consultant is to be ignorant and ask a few questions." Drucker was smart enough to let his questions do the heavy lifting for him.

Like so many of the other techniques we've described in this book, leading by asking questions looks easy. You've been asking questions almost as long as you've been talking, so it's hardly a new concept. Then you try leading by asking, only to find out that what looks easy ... isn't.

Using questions as a leadership tool offers so much leverage it's worth investing some of that valuable asset, your time, to understand questions, and why some questions work far better than others.

TYPES OF QUESTIONS

The function of a question is to elicit a response: "Did you close the valve?" What tells the listener a reply is expected is the word "did," found at the start of the sentence. Beginning a sentence with a word such as do, is, will, should, or how is a common way to form a question.

A second way to form a question is using a triggering word at the end of the sentence: "You closed the valve, correct?" Conversationally, words like right, correct, and ok found at the end of the sentence tell the listener the words are a question, not a statement, and a response is expected.

In some languages, a third way to form a question is by inflection: how the voice sounds when pronouncing the words. "You *closed* the valve?" Reading the words, they would appear to be a sentence; listening to the words in a conversation suggests otherwise. Changing inflection can alter the question and the response: "You closed the *valve*?"

Of course, you do that sort of thing all the time, normally by habit. But habit is unconscious behavior, and the point served by this examination of questions is to create understanding and in turn, facilitate conscious behavior at the point of practice.

Consider the word that signals it's a question as the **Key Word:** that's what triggers the response. The key word can come at the beginning of the question—*did*; it can come at the end of the question: *correct*; it can come subtly in the tone of voice used to express the words: *closed*. If your goal is to have your listener understand what you are looking for in their response, it's obvious that some formatting better serves that purpose than others. You want to make your question understandable—easy—to your follower.

The second important thing to understand is that questions are designed to create one of two kinds of responses. There are questions that will (or should, if the question is properly understood and answered) produce a simple yes or no reply. "Did you close the valve?" The answer is either yes or no. Questions that are intended to produce a yes or no answer are called **Closed Questions**. Closed questions start with words like *do, is, are, will, could, should, have*. There is not a thing wrong with asking a Closed Question—as long as your intention is to get a yes or no answer.

But there are many situations where a yes or no is not what is needed or desired. The person asking the question is looking for more: explanation, details, context, perspective. When that's the kind of reply desired, the question must be asked in a way that produces that information. Questions eliciting that kind of response are called Open-Ended Questions: "Why did you close the valve?" In the reply will be found an explanation.

There are Closed Questions, Open-Ended Questions, and questions that aren't even questions. "What were you thinking?" That's a statement masquerading as a question: "You were not thinking—or thinking properly!" Those are known as rhetorical questions: a question intended to make a point in a dramatic way, not to produce an answer. As a leader, it's best to avoid them entirely: the technique of using questions is demanding enough; no sense making things tougher on you—or your followers.

In this primer for leaders on questions, the third important thing to understand is that Open-Ended Questions can be divided into two types. The first are Direct Questions: these are Open-Ended Questions that come with correct answers: "What is the first step in the procedure to disconnect a hose?" Presumably the answer would be to close the valve. But there can be Open-Ended Questions for which there are not so much correct answers as there are responses that reflect thought, learning, understanding, insight, motivation.

A question that produces effects such as those in the mind of the listener is the stuff of influence!

A DARN GOOD QUESTION

In the States, there's a popular expression dating back more than a century: "That's a Darn Good Question," It has nothing to do with mending socks; it's just a way of saying, "You have me thinking!"—and in a good way. In a nutshell, that is the response a leader wants to trigger when asking a question intended to have influence. Asking a Darn Good Question can be a fabulous way to lead and influence followers.

Like using a hook with a nice juicy piece of bait attached to catch a fish, a Darn Good Question invites participation. Ask, "What are the ways you can think of to determine the tire inflation pressure on a truck?" the listener is very likely to happily answer the question. In the process of answering, all sorts of good things can happen: the listener might apply their current knowledge, evaluate possibilities, and even come up with a way you hadn't thought about. The process of answering the question affects their thinking: they may learn something in the process or reach some new conclusion.

That is the function of the Darn Good Question. Yes, you could explain all of that information to the listener and hope they're paying attention to what you have to say. Better to let them do the talking and you do the listening. Letting the question do the heavy lifting makes perfect sense, but it is clearly a road less travelled.

There are three simple elements that add up to a Darn Good Question.

1. Purpose

The process of leading by asking a Darn Good question starts with having a specific leadership and influence purpose in mind. That is for you to choose. You're not asking a question simply to gather information—or to pester people. There's a point to your question: something to do with influence.

For example, your point might be to get followers to pause, reflect on something important to safety such as risk, and better appreciate the real risks they are taking. That's the purpose behind the question "What's the biggest risk you've taken in the last 24 hours?"

Suppose you come upon someone not following the safety rules, for example by not wearing the required hearing protection. The question "Why aren't you wearing your earplugs?" serves several potential purposes, but the principal one is to get the person to examine the reasons he or she isn't following the rules.

2. Audience

Given the purpose—influence in some way—there is always an appropriate audience for your question.

Some questions can be asked of anyone: the question about biggest risk applies to anyone in the outfit, from the mailroom to the executive officers. Other questions have a far more narrow and specific audience. It might be very appropriate to ask a pipefitter "What do you think the error rate is for the operator who isolated the line you're about to break into?" Or the operator who isolated the equipment, "What are the potential consequences to others if you were to make an error in performing this task?" Or an electrician, "How else might power get to that pump you're going to disconnect, even if you close the breaker?" It doesn't take a lot of imagination to figure out how each of these scenarios might actually get someone hurt. Each has.

As a reminder, you already know the answers to these questions. You could simply tell your followers what the right answers are. Your goal in asking a Darn Good Question is to have them telling you.

Purpose is coupled with the audience: every question has an appropriate audience. And every audience has specific interests and needs. Now to the question.

3. Question

By definition Darn Good Questions are always open-ended: no "yes or no" answers. There's an easy way to guarantee your question is open-ended: the first word of your question must come from this list: who, what, when, where, how or why.

Surely you've heard of these six. Think of any of these six as the **Key Word** in your question.

Limiting yourself to these six **Key Words** simplifies matters for you. Having to start with the **Key Word** will keep you from making a common mistake, asking "Was there a reason why you closed the valve?" Read that question closely: yes it asks why, but the **Key Word** sits in the middle of the question, making it a closed question. It calls for a yes or no answer. Were the answer to be yes, you would certainly want to know what the reason was. That would require asking a second question, "What was your reason?"

There's more: a subtle and powerful genius built into this list of six **Key Words** for Darn Good Questions. It is as if there's a directional arrow of influence for each word, pointing the listener to think about and talk about a specific type of information.

Based on which of the six words you choose as your **Key Word**, the information you will hear in their answer is guaranteed:

- *Who* points to people
- *What* points to thing and specifics
- *When* points to time
- *Where* points to place
- *How* points at means and method
- *Why* takes the audience in the direction of matters like judgment, motivation, and rationale.

The wonderful thing about this directional arrow is that you get to choose the topic, and your followers will surely go in the direction you point them (Figure 9.1). Choose wisely.

When formatting your question, you must pay close and careful attention to the exact words you choose and speak. For most leaders, that's not the usual practice. A slight alternation of the phrasing of a question can change a Direct Question to a Darn Good Question—and vice versa, all the way back to a Closed Question.

For example, the Direct Question, "What is the proper inflation pressure for your tires?" can be changed to a Darn Good Question by asking, "How would you determine the proper inflation pressure for the tires on your car?" But the Darn Good Question can revert to a Closed Question by asking, "Do you know the proper inflation pressure for the tires on your car?"

By choosing a different Key Word, the information provided will change.

THE DIRECTIONAL ARROW
OF THE KEY WORD

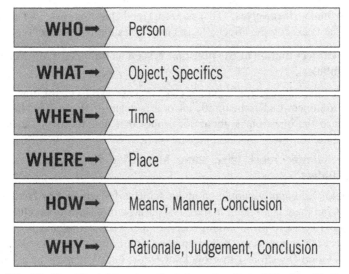

WHO→	Person
WHAT→	Object, Specifics
WHEN→	Time
WHERE→	Place
HOW→	Means, Manner, Conclusion
WHY→	Rationale, Judgement, Conclusion

Figure 9.1   The Key Word determines the information that will be found in the answer.

"How does underinflation affect the performance of a tire?"

"What are the effects of underinflated tires on the handling of a car?"

"Where can you look to find out the proper inflation pressure for the tire pressure on your car?"

While these different variations involve the same subject—tire inflation pressure—asking *what* produces specific details, and *how* focuses on process. Were you to ask in a safety meeting, "Who's ever had to deal with the consequences of driving with underinflated tires?" it would likely cause someone to share what would amount to a testimonial to the value of regularly checking the tires.

That is the stuff of genius.

THE RESPONSE

As simple as asking a Darn Good Question is—it is just an open-ended question starting with the words who, what, when, where, how, or why, asked by a leader to influence a follower—and as powerful as a Darn Good Question can be—picture a fully engaged follower, talking about exactly in the way a leader wants them to—the successful use of the technique is entirely in the hands of the follower. The leader is totally dependent on the follower to answer the question. If the question is greeted by silence, there's a problem.

If nothing else, this serves as a reminder of the Leadership Model defined in the first chapter. The power of followers is rooted in their decision to follow—or not.

On that point, there's much more to triggering a response to a **Darn Good Question** than just the Key Word.

Put yourself in your follower's shoes for a moment: what would cause a follower to think to themselves, "That's a Darn Good Question, so let me tell you what I think." That is exactly the effect a Darn Good Question is intended to produce.

1. **The subject matter of the question is seen as important and relevant – to the follower.**

   Other things being equal, people can't be expected to invest time and energy into a subject that is insignificant or irrelevant to them. Bear in mind, just because the question is about something that matters to the leader doesn't mean the follower will see things the same way.

2. **The follower must have some knowledge by which to understand the subject.**

   People will happily babble on about subjects for which they have no knowledge, but there is no point in that. The purpose of a Darn Good Question is not simply to get a follower talking. On the other hand, complete knowledge of the subject is not necessary, and often not even desired. One of the best uses of a Darn Good Question is to grow knowledge, but there must be a jumping off point to do that.

3. **The follower must consider the question safe to answer.**

   Never underestimate the intelligence of a follower. When they first hear the question, running through their minds will be a different question, "What's going to happen if I answer this question the wrong way?" Followers don't want to self-incriminate, expose their lack of knowledge or be subject to ridicule for a wrong answer. If the follower is the least bit suspicious the question is a trap or a trick, any intelligent follower will evade—or just not answer. Who could blame them?

   An unsafe question is a show-stopper!

4. **The follower needs to see the subject of the question as interesting.**

   Just because the subject is important and relevant does not make it interesting to followers. That's part of the safety leadership challenge: boring questions about important subjects often fail to advance the cause of learning. But there's a huge difference between an interesting subject and an interesting question.

   Consider a question on a life critical safety procedure taught to someone in refresher training, "What are the requirements to enter a confined space?" They know the answer by rote. Compare that with beginning with case where someone died when not following the confined space requirements by asking a Darn Good Question, "When you hear about a fatality involving a confined space, what specific requirement do you think might not have been followed?"

   Given the proper context for the question, just about any subject can become interesting to followers. Context sets up the way to look at a question and can put a subject in a new and different light. "What might be an example of a confined space that most people would not recognize as being one?"

Even with few or no facts, a question can be framed as speculation: "Why do you think a handrail on a stairway would fail when being used?" People love to speculate, so let them. Just make sure the speculation leads to something useful, such as the requirements for your followers to safely enter a confined space or to periodically check the handrails they rely on.

Taken together, these four points establish the criteria by which a follower will judge a Darn Good Question and decide whether or not to volunteer an answer.

- Important and relevant
- Knowledge
- Safe to answer
- Interesting

By understanding these criteria, before you ask your Darn Good Question you can predetermine the probability of a favorable response. If something's missing—for example, your question will come across as unimportant, worse, or unsafe—you can edit your question before you ask.

And if you ask what you think is a Darn Good Question, and hear little in the way of reply, you can use these criteria to figure out why.

PUTTING DARN GOOD QUESTIONS INTO PRACTICE

How your Darn Good Question is phrased can affect the likelihood of getting an answer. Adults don't like to be wrong, particularly in front of the boss and their peers. When they're asked a Direct Question—one with a specific and correct answer—they may be reluctant to volunteer an answer if they aren't sure. Ask, "What are the three steps for asking a Darn Good Question?" you'll hear only from someone who is sure they know the right answer. If nobody responds, the silence can be awkward.

You want to make your questions easier to answer. You can ask, "What are some of the steps to asking a Darn Good Question?" That version doesn't require a complete answer. Asking people for their opinions is safe. So a question starting with "What do you think…" or "In your opinion…" is far more readily addressed than a question calling for a definitive and correct answer. One of the very best questions you can ask near the end of an investigation interview is, "What else can you tell me about the problem that you think would be helpful in understanding what went wrong?" The treasure-load of information that someone is sitting on—but won't volunteer unless asked—might surprise you.

Be patient, with followers—and yourself. Give followers the time to ponder the question and formulate their answers. Being asked questions that aren't aimed at getting information is likely to be a different experience for your followers. Yes, the silence will feel awkward to you. When that happens, avoid the temptation to ask a different question or, worse, answer your own question. In a group, don't "point and shoot" by saying "Paul, what do you think…?" Instead, if there's silence, try counting mentally to 10 before you say anything. If you are sure you have a Darn Good Question, just repeat it. In a group, someone will get the hint.

Asking better questions requires the discipline to pay attention to your words. If you ask, "Does anyone have any questions?" don't be reluctant to correct yourself: "I meant to say 'Who's got a question for me?'"

There's one more important thing to understand to be able to execute this technique well, and it has nothing to do with your followers. To master the technique, you'll probably have to unlearn a lifetime habit—several actually. The first is asking Closed Questions, ones that start with words like *are, would, should, do,* and *will.* Most leaders are not in the habit of asking questions to gain influence, and rarely are leaders practiced in asking questions in the way required by Darn Good Questions.

So, how do you change a habit?

First, change requires recognition of the existing pattern of behavior and the difference between that and what's desired. Next, there needs to be a means of reinforcement, both positive and negative. In practical terms, changing how you use and ask questions boils down to paying close attention to your words. The good news is that with a little bit of practice, and preparation, you'll find asking questions to be one of the simplest—and best—ways to lead ever invented.

Until the technique of asking good, open-ended questions becomes a habit, it's a very good idea to write the questions down ahead of time. Actually, that's a very good way to prepare no matter how skilled you become at asking questions. If you ask the question the wrong way, don't be reluctant to say, "Let me re-phrase my question."

Finally, pay attention to what is being said in response to your questions. This has been called *empathetic listening* but it's nothing more than really paying attention to what your follower has to say. Your follower deserves nothing less. Besides, it's in your own best interest to hear what that person is saying. After all, it's the gold that you are digging for. You want to listen—and look to your followers like you're listening. That's your body language.

When they do start answering your questions, think about the power: your followers are talking about what you want them talking about; you're leading, they're following; and everybody is sharing the workload.

Asking Darn Good Questions will do the heavy lifting for you.

CHAPTER 10

MAKING CHANGE HAPPEN

> Their's not to reason why, Their's but to do and die.
>
> *—Lord Tennyson*

8:15 a.m., Tuesday Morning Staff Meeting

The department's supervisors and specialists are gathered in the conference room, paying rapt attention to their Department Manager, sitting at the head of the conference table.

The manager begins: "I hate to be the bearer of bad news, but our leaders at headquarters have done it again. As if we're not already overloaded with everything else that's going on, now they've decided to add a new rule about how to park company vehicles. We'll be required to back into every parking space. If that's not bad enough, they're telling us we have to put a magnet on the front door reminding us to walk around the vehicle before we get back in to drive a parked vehicle.

"I'm not any happier about this change than you are. Don't ask me what they were thinking: nobody called me up to ask for my opinion before they announced the change. You know that's how things are done around here these days.

"Any questions?"

We've all been on the receiving end of a new policy communicated that way; most leaders will admit at some point, they've made the speech to their followers. As to what happens next, it's totally predictable: resistance. When was the last time you announced a change to a safety policy and were greeted with a resounding cheer and, "Thank you so much for looking out for our safety"?

What you are likely to hear from your followers are complaints, often in the form of questions: "When will they ever stop changing the safety rules? Don't they understand this will make our jobs harder, not easier? Why can't they just leave us alone to do our jobs?"

You often find yourself thinking the same things yourself: leaders in middle management are not immune from these frustrations. The odds are high nobody at the top bothered to ask you what you thought about the impending change before sending it your way. If they had, you might have raised exactly the same questions.

It's tempting to answer the questions from your followers honestly—"I know you are frustrated by all the changes, but the pace of change will only accelerate."

Alive and Well at the End of the Day: The Supervisor's Guide to Managing Safety in Operations, Second Edition. Paul D. Balmert.

"Yes, new policies do add to the complexity of the work we do and often take more time." In the future, it's not likely that decisions about how we do our work will be left up to us. Telling followers that truth—which those answers most likely are—will only serve to make matters worse. But telling them otherwise can ruin your credibility.

When it comes to communicating change, it's easy to feel like you're leading the Charge of the Light Brigade; your best hope is to survive the battle.

Change isn't going to stop, and neither is resistance. Effectively communicating change when it comes in the form of new or revised safety procedures is a very important task a leader is responsible for, making it essential that every leader masters the process of "making change happen."

Doing that successfully requires a thorough understanding of the process of change and the best practices to follow.

THREE THINGS TO UNDERSTAND ABOUT CHANGE

Before laying out the details of "What to do" and "How to do that" it's important to appreciate what's really going on in the context of communicating change when the change involves a safety policy, new or revised. On that point there are three things every leader needs to understand: the first is obvious; the second becomes obvious with a moment's thought; the third might come as a surprise.

When there's a proposal to adopt a new safety policy, or modify one that's in place, the odds are high it's a direct response to misfortune. Track every single safety policy and procedure found in any operation function to its point of origin—when rules about lock out, fall protection, process safety, personal protective equipment, using mobile phones and even how to park were first written—you'll find a tragedy. Someone's blood was spilled somewhere on the planet. The intent of every safety policy—and overarching goal of every change—is to prevent that from happening again—here.

That you know, but as noble as that purpose is, it won't stop people from complaining about change.

Resistance to change can come about for a long list of reasons. But be careful not to generalize them, like saying "change requires more work." When the change outlaws doing something—like restricting who is allowed to operate high voltage switchgear—it means doing less work, not more. But as it relates to resistance to a new or revised safety procedure, there is one reason that's guaranteed: the natural human resistance to a change *imposed* on them by others.

Unless it was the follower's idea, a change in a safety policy is imposed on followers by the leader who approved the change. Yes, that leader is convinced the change is a good thing, done to benefit safety. But for the leader's followers, it doesn't change the fact that it's an imposed change. There is no choice but to change, and change to do what the leader wants (Figure 10.1).

That's reality, and that you need to understand. Instead of seeing resistance as a bad thing, you're better off recognizing resistance as a normal and predictable part of the change process, and be prepared to deal with it.

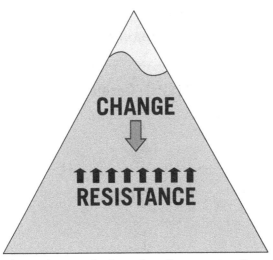

Figure 10.1   The imposed nature of a change in policy guarantees there will be resistance.

The third thing to understand becomes clear when looking at a change in policy as a work process. Remember, all work is a process: a process converts inputs into outputs. In process of making change happen, there are three steps that matter.

The first step starts with identifying the need for a new or revised procedure. That's a big step, often involving smaller steps, like getting input from the process experts and conducting a pilot test of the change to see how it works. This first step ends with a proposal ready to be submitted to the proper level of management for approval. Until approved, a proposal is nothing more than a piece of paper.

Approval is the second step in the process. Leaders, particularly executives, typically see the second step in the process—making the decision—as the most important one in the process. For them, it seems so because it is the step that is their's to perform. Decisions can be tough and unpopular.

But decision made, an approved policy is still nothing more than a piece of paper, now with a leader's signature on it. What happens next is by far and away the most important step in making change happen: execution. See Figure 10.2.

The third step—execution—really is the only step in the change process that matters. Execution is where the change happens: when what's written on a piece of paper is converted into action.

A change in a policy or procedure will change how the work gets done. If it doesn't change how things get done, there is no need to even communicate it. A new policy or procedure becomes reality only when there is a change in the way work is done, and that will involve how people behave in some way.

Behavior determines the extent and success of the change. If the change doesn't happen the way it is supposed to, it makes the policy nothing more than one more piece of paper. An operation that does not follow its own safety procedures will be worse off, not better.

THE WORK PROCESS
TO MAKE CHANGE HAPPEN

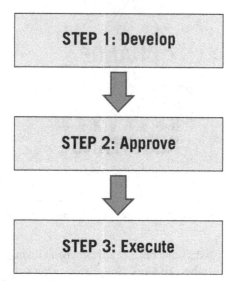

Figure 10.2 Of the three steps involved in changing a procedure, the final step, execution, is by far the most important.

Making change happen is serious business. The duty of executing change falls to the leaders, often the front-line supervisor. That leader is responsible for successfully changing how the work gets done, how people behave, and managing their resistance to what they will see as an imposed change.

Now you fully appreciate the context of change: why getting followers not just to accept change, but actually to make the change, is such a tough challenge.

TWO KEY STEPS

When communicating a change in policy, the end game is to achieve *compliance*. Communication of change over, your followers head out to the job understanding what the change is and committed to make it happen. That's execution. The closer you can come to that goal, the easier your job becomes. The farther away from that goal your followers are, the tougher your job becomes—because your job is to make the change happen.

As to what to do to achieve buy-in and compliance, there are the basic requirements to explain the change and take the questions. But there are two key steps to take that will dramatically up the odds your followers will understand, accept and make the change:

1. Explain the reasons for the change.
2. Determine how compliance with the new requirements will be accomplished.

As to why these two steps make such a difference, the logic for the first step is obvious: adults are far more likely to go along with a change if they understand the reasons for it.

Consider that very good news: Beginning by explaining the reason for the change will always help to reduce resistance to the change and increase buy-in to make the change. When the change involves a safety policy, there's a natural built-in advantage: that reason will invariably be found in an event that produced harm. The purpose of the change is to prevent that same harm from happening again here, to your good followers.

The second key step focuses on the heart of making the change happen: compliance with the change. That the change is communicated to your followers means the work of your followers will be changed in some way. Exactly what way? What has to be done to enable that change? That's for you to determine.

Your might think every change should be accompanied with all the answer to every question about doing that. Rarely will that be the case; in practice, you're better off being left on your own to figure out the best answers for you and your followers. You don't really want that much help from the outside.

It's all part of your job of making change happen.

TEMPLATE FOR MAKING CHANGE HAPPEN

To summarize the principles that apply to changes in safety policies and procedures:

- Changes to a safety procedure start with a negative event; the goal is to keep it from happening again—here.
- Resistance is in part the byproduct of the imposed nature of the change; control—the ability to determine the outcome – does not belong to the followers.
- Execution—making the change happen—is the most important step in the change process; getting followers to change is the essence of the challenge.

With this understanding of the process, the proper way to put these two key steps into practice to make change happen successfully becomes a simple and straightforward matter.

1. Find out the reasons for the change.
2. Identify the issues and questions likely to be raised by your followers.
3. Know how to properly begin the explanation of the change.
4. Know how to "close the sale" gaining the commitment of your followers to make the change.

Consider these four points your **Template for Making Change Happen**. When it's time for you to communicate a change to your followers, you'll want to use these four points as your way of preparing for and communicating the change.

Point 1: The way to overcome resistance starts by finding out the reason for the change. Then you want to explain the reason for the change before communicating the change. If the vehicle parking policy was changed because of a rash of accidents involving backing out—including one near-fatal collision with a pedestrian—explaining that first increases the likelihood your followers will go along with the change.

Often the reason for the change isn't provided as part of the policy. That's a shame, as there is great value in documenting the reason in the policy itself. Trevor Kletz, a historian of accidents in the process industry, found the same accidents tended to occur in roughly 25-year cycles. What's special about 25 years? That's about how long it took for everyone who was there at the time—and learned the lesson painfully—to have moved on. That leaves the next generation to learn the same lesson the hard way. Kletz said: "Organizations don't have memories; people do." That's the benefit of writing the reason for the change in the policy or procedure.

When the new procedure doesn't specify its reason, invest a few minutes finding out. The explanation is seldom a secret, or all that complicated. Sometimes executives assume the reason is obvious and doesn't need to be communicated. Don't fall for that assumption: find out.

Point 2: Resistance becomes easier to deal with when you're prepared for it. You now understand that when policies are changed, resistance is created because the change is imposed by the leaders. Resistance reflects the fleeting hope that if there are enough complaints about a change, leaders will change their collective mind. The odds of that may be small, of course, but that won't stop people from complaining.

On the other hand, a change in policy means that behavior must change, and there are usually issues involved in going from old behavior to new. Raising a *real issue* has nothing in common with *resistance*. Resistance is about sticking with the past; real issues involve dealing with what's going to happen in the future, as the change is put into practice.

The way a real issue is raised might sound like resistance, particularly when your follower embellishes the point with a reference to the lack of intelligence of those approving the change. Ignore the histrionics; pay attention to the point.

Think of getting real issues as good news. When your followers are raising real issues, it's the tip-off to you as the leader they've stepped into the future. Not only that, but they are starting to help you with execution: they're telling you about problems that will need to be dealt with, sooner or later. Sooner is better.

In most cases, what you will get from your followers is a mix of resistance and real issues. You need to be prepared for both—before starting the communication process. Make a list and divide it into categories: resistance and real issues. That way, when the point is raised, you'll know what you're getting.

You know your followers; likely you can predict in advance what they'll say in response to the change. Give a moment's thought to their likely objections and issues: often there is a ready and logical response. To those who say, "This change will cost us money and customers" the reply is to remind them of the Case for Safety: "Tell me one business objective we have that is so important it's worth someone getting seriously hurt—or worse."

The first two points of the template—knowing the reason for the change and identifying both resistance and real issues—involve preparation. A small amount of preparation will go a long way toward creating buy-in and execution.

Time to begin the communication.

Point 3: Starting the policy change communication with the disclaimer that "Our leaders are up to their usual tricks" might seem like an honest admission by a leader proud of their reputation for openness. But telling followers that you're not the least bit in favor of a change does nothing to cause compliance. It's more likely to do the opposite—increase resistance—making your job of execution even more difficult. So, as tempting as it might be to criticize the policy, you're far better off keeping those opinions to yourself.

Of course, when you know the reason the change was made, the odds increase that you'll find yourself supporting it. Starting by explaining why the change was made is a much smarter strategy: "You might have heard we've suffered a rash of vehicle incidents involving backing out of parking spaces. Two weeks ago, at another site, a pedestrian was struck by a pickup backing out. The driver said they looked every way before and didn't see anyone in harm's way. The person struck suffered life-threatening injuries. Because of these incidents, we are going to change our policy about parking company vehicles."

An alternative approach would be to start with one of those Darn Good Questions. There's a purpose to the question, and that purpose isn't about getting information; it's about influencing.

In the case of a new policy, that purpose would be to create support for the need to change the policy. In creating support for the parking policy, a Darn Good Question to ask might go, "What's been your experience when you've been walking through a parking area, where other drivers are backing out of their parking spaces?" A question like that is bound to stimulate discussion, and the personal experiences shared will likely demonstrate the need for change. That's letting the Darn Good Question do the heavy lifting for you.

Point 4: Explanation given, policy communicated, objections heard, and issues raised, it's time to close the sale. Remember: your goal is execution, which means all your followers leave the meeting understanding what is required and committed to change. What do you say to accomplish that?

Going into the meeting knowing your close is always a good practice; the greater the resistance expected, the more valuable that preparation. You can close with a reminder about safety in the form of your Stump Speech: "I know I've said it many times before, but nothing is more important than going home safely. We get reminded of that when we hear about an accident like the one that just happened." You can remind everyone that compliance is expected and that you'll be complying just like everyone else. That's nothing more than Leading by Example. Acknowledging the impact of the change on how the work gets done helps to set the stage for dealing with the unexpected impacts of the change as they become evident. "It's going to take some getting used to. And we'll need to make sure there's an ample supply of magnets. We all know how things like that can disappear."

Finally, there is the matter of your personal support for the change. Over the years many leaders have taken a neutral stance on their personal opinions: "As the supervisor, I wasn't party to the decision, and everyone knows that." It might seem when the leader tells followers when they're in favor of a change, they're obligated to do the same when they're not. All the more reason to keep quiet: "The Company has decided to…"

That approach misses a huge opportunity to cash in on your influence. Studies consistently show that of all the levels of management, the one with the most credibility is the follower's immediate front-line supervisor. To the crew, what the supervisor thinks matters significantly. In the case of safety policies, more often than not, as the supervisor, you are of the opinion the change is a good one.

So tell them what you think.

Doing so might just bring some of the undecideds over to your side. In those few situations in which you don't happen to agree with the change, keep that opinion to yourself. You can just tell your people, "What I think about the new policy doesn't really matter. Now, what do we need to do to make the new policy work?" They'll figure out that you're not really excited about the change, but you won't have planted any seeds of doubt about your commitment to following though.

The better question is always "What do we need to do to make the new policy work?" It focuses on the future—life under the new policy. Asking it is a great way to break a logjam of resistance, one in which people are stuck on thinking about the reasons why not to make the change, rather than how to make the change happen. Solicit the ideas and issues and assign follow-up where needed. Need to stock new forms, get additional approvals before working, conduct an additional check? What equipment do we need? Who needs to be trained? When?

Many new policies are accompanied by the boilerplate "Failure to follow the policy can result in corrective action, up to and including termination." It's only fair to let everyone know the truth: compliance is required, and the failure to comply comes with potential consequences. Some of those consequences are administrative in nature: verbal warning, letter of reprimand, time off, or even the ultimate administrative consequence, termination for cause. Of course, those consequences exist only to the extent that management is present and committed to enforce the policy.

Consider the other potential consequence in play when the safety rules aren't followed. Picture the failure to comply results in a serious injury, or worse, a fatality: for example, a driver hits a pedestrian when backing out of a parking space. By far, that is more significant than an administrative consequence.

Effectively communicating the administrative consequences is really a matter of perspective. Yes, someone can get in trouble if found in violation of the policy. But better to follow the policy and avoid the pain and suffering from the event.

EXECUTION, EXECUTION, EXECUTION!

Making change happen is easier and better when your followers have bought into the new policy. That requires you to be in the selling mode; you'd rather do that than to spend time in the enforcement mode.

By following these commonsense practices, the communication of change will increase the probability of creating understanding, acceptance and compliance with the change:

A) Explain the reasons for the change.

  • Find out the reason for the change.

- Explain the reason first; the change second.
- See resistance as a normal part of the change process.

B)   Determine how compliance with the new requirements will be accomplished.

- Remember when policies change, work must change.
- Determine the impact of the change on how work gets done.
- Solve the problems necessary to enable the change.
- Create an incentive to comply both through words—with vocal support—and through actions—dealing with the real issues and leading by example.

Finally, never forget your leaders are depending on you to implement the change successfully and completely. That's why your role in making change happen is the most important of any one leader in the operation.

CHAPTER 11

UNDERSTANDING WHAT WENT WRONG

I learned something out there today. I'm just not sure what it was.

—*John McEnroe*

Every day, things don't go exactly as planned. Equipment breaks down; production gets delayed; schedules aren't met; work costs more than was budgeted; customers call with complaints. The high level of activity going on in operations, plus the probability of human error and equipment failure, make the likelihood of encountering problems high. Thought of in that light, it's amazing there aren't even more problems for a leader to face.

Sometimes the problems involve safety. Here's one example.

7:15 a.m., Toolbox Safety Meeting

The supervisor is in the final stretch of his morning meeting: "OK. Anything else to share on the safety front?"

Jim, a crew member, volunteers a "safety moment." "Hate to admit this, but I had a little near miss yesterday. I borrowed one of the bikes to pick up some product samples from the lab. Coming back to our building, I rode off the edge of the sidewalk. It's pretty narrow there, and there isn't a guardrail. I'd recommend in the future being extra careful when riding there."

The supervisor's first interest is with Jim's safety: "Jim, thanks for sharing that incident. At some point, just about every one of us has had an accident with a bike. Are you OK?"

Jim smiles ruefully. "Oh, yeah. No problem with me, but the samples spilled out in the mud. And the tire went flat when I went off the edge. I had to borrow the other bike and ride back up to the lab to pick up another set of samples."

The supervisor puts a wrap on the discussion: "As Jim points out, even something as simple as riding a bike can result in an accident. So be careful on those routine duties. Let's go to work, and work safely."

After the room clears, the supervisor reflects on the incident. "Stuff like that goes on all the time. What should I do about what I just heard?"

Alive and Well at the End of the Day: The Supervisor's Guide to Managing Safety in Operations, Second Edition. Paul D. Balmert.
© 2023 John Wiley & Sons, Inc. Published 2023 by John Wiley & Sons, Inc.

It seems like such a minor problem, just one more of those many small moments in the life of a leader. Realistically, the odds of launching a formal root cause investigation for an incident like this aren't very high. But as with so many seemingly minor events, when things go wrong, there's a lot more going on than first meets the eye.

IGNORE OR INVESTIGATE?

The first thing to understand about problems is that every time something goes wrong, a Moment of High Influence is created. A failure, big or small, is one more of those everyday events during which followers are in a high state of readiness to be influenced, paying attention to what their leaders say and do in the aftermath of the failure. That means a failure represents an opportunity to stand up and lead.

Fortunately, most of the time problems are small and, with a little bit of attention on the part of the supervisor, easily fixed. Thanks to the practice they get at this part of the job, leaders are normally very good at fixing problems and moving on. But there is the related question: when does it become necessary to dig into the cause of a problem?

If the problem is big enough to show up on the radar screen of top management—a serious injury, a significant production interruption, a dissatisfied customer, substantial property damage, a major process event—a formal investigation is usually required by policy. When the event gets reported, the investigation is commissioned. These formal investigations take time and skill, the stakes are high, and they are often led by someone with expertise in a specific investigation method. The formal investigation process may be great for getting to the bottom of a major incident, but, as a supervisor or manager, you hope something like that is not ever required on your watch.

Still, whether large or small, every problem has its causes. What do you do about all the relatively minor problems you face every day? The ones that don't go beyond your desk? Is it enough to deal with the consequences—fix the problem—and ignore the causes?

Would you investigate a problem like this?

9:42 a.m. TBD Unit

As the dropped pipe wrench hurtled to the earth, the pipefitter screamed, "Look out below." Fortunately, Mother Earth took the hit for the team, as the dropped object narrowly missed three people working at ground level.

The wrench was unharmed.

It's called a near miss: as if the pipefitter had been aiming the pipe wrench at the enemy below—and missed. A close call would be a better description. From personal experience, we know there are far more close calls than there are real events with consequences.

THE INJURY TRIANGLE

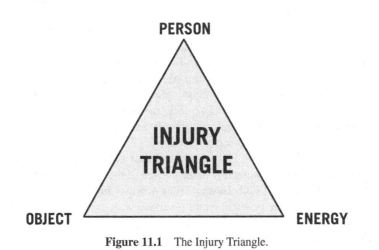

Figure 11.1   The Injury Triangle.

People are hard targets to hit, so the odds favor misses over hits. If your followers have suffered a few injuries, it logically follows that they have had many more near misses. That takes us back to the Injury Triangle. See Figure 11.1.

How many of those near misses have you heard about?

Many near misses go unreported. It's not hard to figure out why: reporting leads to investigation. Conduct an investigation, somebody is likely to be found to be the cause. Be that cause, it's not a good thing. Not reporting near misses is always the path of least resistance. But not reporting a near miss means that the same thing could happen again because the source of the problem goes undetermined and the problem unfixed. Every time that happens, there can be another opportunity for a hit instead of a miss.

Report, and something bad might happen. Don't report, and something bad might happen. This is not an easy situation for either followers or their leader.

If you're the leader, when a follower reports a near miss to you, the first thing to understand is that you've been given a vote of confidence. They trust you enough to tell you what went wrong—particularly if nobody saw what happened, like our follower at the tool box safety meeting who reported what was described as an unwitnessed "near miss." No, it wasn't really a near miss: there was an event producing damage to both a tire and product samples. And yes, he should have reported it when it happened—yesterday. A supervisor might be tempted to come down on the side of those offenses. Tread very carefully: that may well be a case of being "too clever by half." The follower didn't have to tell the supervisor, and many in that situation wouldn't. If chastised, the follower may never report another unwitnessed near-miss. Worse, they will share their unfortunate experience to everyone they know: "What was I thinking when I told the boss about that near miss?" More near misses will be driven underground.

Back to our question: ignore or investigate? Most problems are small; when is it sufficient to fix the problem and move on, and when is it necessary to find out what went wrong and determine an appropriate solution?

First, many of the problems you see are minor only because the consequences resulting from the situation played out that way. Change the facts just a bit and the consequences might well have been far different: the dropped pipe wrench could have hit and killed one of those working below; a dropped load could have fallen 10 feet instead of 10 inches; the wrong sample could have been sent to the biggest customer; instead of springing a pinhole leak, the pipe could have burst. The difference between a minor problem and a catastrophe often is a matter of time and place.

Said another way, that's just plain luck.

Second, small problems represent wonderful opportunities to learn what is really going on in your operation. Fix at least some of them, and your life might actually get easier. If you don't find out what went wrong—and do something about it— you can't expect the same problem won't resurface with even bigger consequences.

Third, when a follower reports a near miss, they expect you to do something about the situation. A near-miss report creates a huge Moment of High Influence. It's an opportunity to make things better without anyone or anything suffering any harm. Consider these close calls a message: fix the problem!

When you're investigating small problems and near misses, you won't find yourself under outside pressure to get to the bottom of things and to administer some form of corrective action to the "guilty parties." You shouldn't be facing the level of apprehension on the part of your followers that you'd see after a serious event.

In a perfect world, you would always find the causes of every problem you did not want to happen again. But you don't work there; as a practical matter there are choices to be made—by you. The first choice is to decide which incidents warrant your time and attention—beyond simply dealing with the consequences and fixing the problem.

Deciding a problem is worth digging into doesn't have to cause the commissioning a full-blown root cause investigation, calling in the parties in for a meeting in a conference room, and publishing a report. Over time, you'll be much better off looking into more problems in an informal way and understanding what went wrong.

There's a simple and fast way to do exactly that.

ABOUT FAULT

Quality guru Phillip Crosby established the principle "Find the root cause and fix it," as one of his Four Absolutes of Quality. Doing that makes perfect sense: problems always have causes. If you fix the root cause, the problem goes away. Permanently.

Finding the root cause sounds simple. In theory it is. But don't confuse simple with easy. In practice, getting to the truth about what went wrong is a tough challenge. Lead an accident investigation and you're usually looking at a room full of very nervous followers who are anything but anxious to see to it that the whole truth is uncovered about what went wrong.

The source of that tension is found in the Investigation Dilemma. The roots of almost every failure are found in human behavior: there will be fingerprints on the causes. Everybody knows that. Everyone has a good idea as to what will happen to those found responsible for the failure. That means an accident and its subsequent investigation will feel like crime and punishment; which explains why people often claim, "It wasn't my fault." And why discovering the truth often proves difficult.

That dysfunctional tension has led many leaders to the conclusion that if fear could be eliminated, there wouldn't be any problem in finding the truth. Hence the advice, "Fix the problem, not the blame."

That's what led to the creation of many of the popular root cause investigation methods. Logic diagrams, flow charts, computer programs, and sophisticated terminology have become parts of the investigation landscape. Following a systematic method never hurts, particularly when the person leading the investigation is a skillful practitioner.

But don't think for a moment that following one of those methods drives out tension and guarantees you'll find out what really went wrong. When you read root cause with a label like "management system failure" or "equipment design" it's an indication that the method has only succeeded in making the investigation report look better on paper.

There's no sense pretending tension doesn't exist in your search for the truth as to what went wrong. That discomfort can exist anywhere. In Midland Texas, a truck driver drove a parade float with twenty-four people through a railroad crossing with the lights flashing, the gates coming down, and a train was approaching at 62 miles per hour with its horn blasting. None of those warnings stopped the driver from heading into harm's way. The train struck the float, killing four; among the collateral damage, the parade trailer struck the vehicle driven by the police escort. The National Transportation Safety Board investigated this horrible tragedy, noting "The primary responsibility of a motor vehicle operator…is to operate the vehicle safely. The responsibility rests solely with the driver of the vehicle."

Those facts and requirements did not stop an independent Board from determining the root cause to be the "failure of the city of Midland…to identify and mitigate the risks associated with routing a parade through a highway-railroad grade crossing" and, because there was a police escort, create "an expectancy of safety on the part of the float driver."

The truth is that it's easy to find fault with institutions, systems and processes and faceless groups and tough to pinpoint an individual as causing the failure.

As the leader responsible for sending everyone home, alive and well, you need a better way to think about the tension that's created when things go wrong. Picture a world of work where there is no fear of consequences for doing the wrong thing. Sure, people wouldn't be the least bit reluctant to own up to what they did wrong, making the job of finding out what went wrong easy. But with no fear of consequences, why would people do the right thing in the first place? Without consequences, why follow the rules—particularly those that are inconvenient? Why take the time to be careful? Why double-check before acting?

The reality is that the fear of consequences is what often drives people to do the right thing in the first place. Take that fear away, there's no telling how many things would be done wrong.

Never lose sight of the fact that there are two consequences in play when people take risk and don't work safely: the consequence of getting into trouble, and the consequence of getting hurt. It's the latter consequence everyone should fear the most. That's the Case for Safety, one more time.

With that perspective, the tension in the aftermath of failure is actually a good thing. It just needs to be managed better.

LEADING IN THE MOMENT

Understanding what causes tension when things go wrong won't make it go away. As a leader you're still left with the matter of how to manage the tension; this is always part of the process of understanding what went wrong. Good leadership helps. If your goal is to get as close to the whole truth as possible, there are steps you can take to lead better when things go wrong.

Henry Ford said, "Failure is the opportunity to begin again—more intelligently." Failure also creates a Moment of High Influence. Something undesirable has happened. In the case of safety, that something might be an event that got someone hurt or equipment damaged, or nearly did so. In either case, people are sitting up and paying attention to what their leaders say and do.

In the aftermath of a failure your actions will speak more powerfully than words, but your words matter. Having the right words to fit this situation is vital. You need to know what to say to the person who's been injured, and to those who might be present at the meeting and interviewed about what went wrong. Many leaders think the situation speaks for itself. It doesn't. So, before the next failure surfaces, invest some time thinking about your beliefs and expectations about failure, injuries, what to do about them.

That investment of your time might lead you to this view about your investigation practice:

> You're looking into the cause of an event, one that you wish hadn't happened, and one that you will work very hard to prevent from happening. If it caused someone to get hurt, you truly regret it. Keeping people safe is your most important goal as a leader.

> There's the matter of perspective. No matter how bad things seem, every event could have been worse.

> But you don't want there to be a next time. That's why you're taking the time to understand what went wrong. Doing that requires honesty: everyone needs to contribute everything he or she knows about the cause of the failure so that it can be prevented from happening again. Sure, doing that is never easy. Nobody likes to be wrong or to be responsible for doing the wrong thing that caused harm.

> What's the alternative? Blaming the system? Or some object? Or God? Do that, and everyone feels good—until the next time.

If that's what you're thinking, you would do well to share your view with your followers. Think of it as your Stump Speech for an investigation. Here's an example of how that might go:

> We're here looking into the cause of an incident. We wish this hadn't happened, and we all work very hard to prevent things like this from happening. Keeping people safe is my most important goal as a leader.

> But let's keep the proper perspective here. No matter how bad things seem, this could have been worse.

> We don't want there to be a next time. That's why we're taking the time to understand what went wrong.

> Doing that requires honesty: everyone needs to contribute everything they know about the cause of the failure so that it can be prevented from happening again. Sure, doing that is never easy. Nobody likes to be wrong or to be responsible for doing the wrong thing.

> What's the alternative? Blaming the system? Some object? An act of God?

> Do that, and everyone feels good—until the next time, when the same failure happens. Then, we'll all feel terrible. And responsible.

Words like those, coming from a leader like you, will help ease the tension and up the odds you get to the truth.

When a near miss is reported, you can ask what might be considered a Darn Good Question: "How often have we had a close call like this one?" You might be surprised to learn that this is not the first time the problem has happened. You may well be living on borrowed time.

Don't miss out on the opportunity to recognize the person who reported the incident in the first place. The act of reporting is a Moment of High Influence and the perfect opportunity to recognize and reinforce the right behavior. The SORRY positive model represents the best practice to do that.

But, since actions speak louder than words, your followers will be looking for what you will do to correct the problem.

THE FUNDAMENTAL QUESTIONS

Your operation may utilize a standard root cause methodology to perform investigations, and you may have been trained in how to follow this process, and skillful at doing that. If so, use the process. But most organizations don't, and for the ones that do, when it comes time to use a prescribed method to investigate a problem, many supervisors and managers struggle. Proficiency requires a lot of practice.

There is a simple and practical alternative to any of these formal methods, one so easy to use that it requires little practice. Following the technique will likely find out what a supervisor needs to understand about simple, everyday problems. Using the process requires no formal training, doesn't obligate holding a meeting, and it won't take a lot of time.

As to the method, it's based on the function of an investigation into a problem: any problem, not just safety problems: that is to understand what went wrong. It seems obvious, but it's worth stating to provide focus.

As to corrective action—fixing the problem—it's certainly part of the investigation process. Actually, it is the reason for the investigation. But without correctly understanding what went wrong, any planned corrective action is a roll of the dice.

When something goes wrong, what needs to be understood about the failure? The answer seems equally obvious: *why* it happened.

There's a popular theory of finding root cause that suggests all that is needed is to keep asking the question *why* enough times to get to the bottom of things. You might. Try asking someone who slipped and fell on an icy sidewalk the question *why*: "Because ice is very slippery." If you're next *why is*: "Why is ice so slippery" you're not on a good path to gain any useful understanding about why the person fell. Heading in a different direction might not be all that productive, either. "Why did you walk on the ice-covered sidewalk?" might get you, "That was the route to the parking lot."

The approach of asking *why* repetitively glosses over a vitally important point in understanding what went wrong: it doesn't explain *how* something happened. Had the ice been sanded? What kind of footwear was worn? Was there a handrail that could have been held? Was it? Was there an alternative route to the parking lot that wasn't icy?

In practical terms, the function of an investigation is to understand *how* an event happened and *why* it happened. Based on that understanding, corrective action can be determined.

As you now understand from reading Chapter 9, "The Power of Questions", the key words *how* and *why* produce entirely different types of information. It's easy to confuse *how* with *why*. After all, they both describe factors that combine to cause something to happen. But what they describe are completely different: *how* explains process: the means, manner, and method; *why* explains reason, rationale, and conclusion.

Here's an example from the world of physics that illustrates the difference between *how* and *why*:

Drop a stone from a tower and it falls to the ground. You can measure the direction, speed, and distance of its travel. That is *how* it falls. How provides the answers in the form of facts and data.

The answer to the question "Why does a stone fall down?" puzzled philosophers for thousands of years. Finally, Isaac Newton provided an explanation that satisfied everyone, at least for a few hundred years: gravity.

Why explains the fact that the stone falls down and not up or sideways. Gravity is the answer to the question "Why does the stone fall?" *Why* explains the reason for the behavior of the object, since the reason for that behavior is not obvious to the eye. *How* and *why* are equally important questions to both ask and understand. It is true for science, and equally true for investigating problems.

How did the load fall? "The load dropped when the chain fall broke. The chain fall broke because the load exceeded its capacity by 500 pounds."

Why did the load fall? "We conclude the reason the chain fall was overloaded, was that the load was marked as weighing half of its actual weight. The calculation of the load weight was incorrect." Of course, that begs the question, *"How* did that happen?"

The distinction between *why* something happened and *how* it happened is essential to understanding what went wrong. You need good answers to both questions. Combined, how and why are the essence of an investigation into any problem.

Digging deeper into the information needed to understand *how* leads to the predictable need for additional information: answers to *when, where, what, and who.* This should begin to sound familiar. It's what every journalism major is taught to report in a newspaper story, and what was explained as the potential **Key Words** for a Darn Good Question (Figure 11.2).

Taken together the answers to these six questions—who, what, when, where, how, and why—represent the fundamental information determined by an investigation. In this application, think of these as the **Fundamental Questions**.

There is an additional piece of information to be discovered: the damage caused by the event. Those are the *consequences* of what went wrong.

But if the event were a near miss, there would have been no consequences; that is what makes it a near miss instead of a hit. That being the case, consequences need to be looked at in two ways: the *actual* and *potential* consequences. The dropped pipe wrench caused no harm, but it had the potential to have caused a fatality.

In terms of organizing the answers to six questions and the damage—actual and potential—there is a simple way to do that: The **Investigation T Chart**. See Figure 11.3.

The divide of the information in the two columns on the T Chart is noteworthy: on the left side are found facts; facts can be proven true. Actual consequences fit that definition, too. You know what the consequences were.

THE FUNDAMENTAL QUESTIONS

WHO→	Person
WHAT→	Objects, Specifics
WHEN→	Time
WHERE→	Place
HOW→	Means, Manor, Conclusion
WHY→	Rationale, Judgment, Conclusion

Figure 11.2   The Key Words establish the answers to the Fundamental Questions, explaining what went wrong.

THE INVESTIGATION T CHART

Who/What/When/Where/How (facts, data, evidence)	Why (conclusion)

Who: Jim
 	
What: Bike
 	
When: Yesterday
 	
Where: Sidewalk Building Entrance
 	
How: Riding bike
  Slow speed
  Bike went into ditch
  Holding handlebar
  with one hand
	Why: Picking up Samples
	  Sidewalk shortest route
	  No basket on bike
	  No handrail
	  Common practice

Consequences – Actual • *Samples spilled* • *Flat tire*	*Consequences – Potential* • *Serious injury from fall*

Figure 11.3   The T Chart is a simple way to organize the answers to these questions, putting facts in the left column and opinions and conclusions in the right column.

Often facts are obvious because there is ample evidence to prove them. "He tripped and fell down the flight of stairs. Three people watched him as he fell and he was lying at the foot of the stairs when the ambulance arrived." Sometimes facts are not obvious: "How long had this situation existed before someone recognized there was a problem?" You may have to dig deeper, and perhaps at some point admit, 'I don't know.'

When you find out everything there is to know, the answers to the questions *who, what, when, where,* and *how* will always be found in facts, data, and evidence. By comparison, answers to *why* reflect a conclusion based on the facts. Answers to why reflect the judgment and opinion of others: "He wasn't paying enough attention while he was walking down the stairs." It's what most people mean when the term "root cause" is used. Properly reached, a conclusion must be based on the facts.

Similarly, potential consequences are a matter of opinion. In most situations, the potential consequences from what could have happened will be greater than the actual consequences from what happened. Said another way, no matter how bad the event was, it could have been worse.

If you understand the consequences—actual and potential—they will guide you to make the best choice as to whether to ignore or investigate.

As to putting this technique into practice—asking the Fundamental Questions and gathering information using the T Chart—it's easy.

At the point when you first learn about a problem, start with the consequences: was the damage—actual and potential—significant enough and warrant looking into the cause? If they are, take out a piece of paper and draw a line down the middle. Put the answers to who, what, when, where, and how on the left side: those are the facts. Put your conclusions on the right.

You don't have to ask all the questions and get all the answers in one setting.

THE PROBLEM WITH SOLUTIONS

Most of us think the difficult part of the investigation is over once we've found the cause. As challenging as that task might be, finding out what went wrong is only half the battle. The other half—fixing the problem—isn't any easier. But it is the point of the investigation: if the problem isn't fixed, there is no benefit from investigating it.

That makes perfect sense, but when you read through the corrective action commitments that typically follow an investigation, you'll see an all-too-familiar list:

- "Replace the defective part."
- "Repair the equipment."
- "Remind everyone to follow the procedure."
- "Share the lessons learned with the rest of the organization."
- "Retrain the person involved."
- "Revise and reissue the procedure."

Did anyone really think "solutions" like these will solve anything?

There's an adage, "The definition of insanity is doing the same thing over and over expecting a different result." So-called solutions like those are regularly proposed on investigations; moreover, those who review the report often approve them. It's as if everyone's going through the motions, checking the boxes that say, "investigated" and "corrective action taken."

It doesn't get any better when there's a major investigation led by outside experts. Consider the problem of people not following the safety rules—not one or two people, but an entire organization. In that case, rules are thought of as a suggestion rather than a requirement. More than one major investigation into this kind of problem reached the conclusion that the culture was broken and needed to be fixed.

Yes, the culture was part of the problem. Given that problem, the solution would be to fix the culture. But suggesting that is about as useful as recommending world peace as the solution to war.

Fixing the problem calls for some honest thinking on the part of the leader. First about the causes: are they really the correct explanation for what went wrong? If not, better to go back to square one and get the real answers to the Fundamental Questions than to roll the dice.

If the answers are correct, it's time to face reality about what it takes to actually solve the problem. For example:

- If someone knew how to do something and didn't do it, training is not a solution.
- If the procedure is adequate but not followed, rewriting or reissuing it won't change the level of compliance.
- Reminding people to "be careful" seldom makes anyone take additional care.
- Sharing "lessons learned" doesn't guarantee they are learned and applied.
- If the cause involves behavior, fixing the object that caused the injury does nothing to change the underlying behavior.

Not every problem has an easy and straightforward solution. The tough ones call for a leader to think better—and harder—about what it will really take to fix the problem, particularly when it comes to behavior. Don't fall victim to thinking that putting what the firefighters call "political water on a fire"—a solution that gives the appearance of doing something, even if it does no good—will suffice.

If it doesn't and there's a more serious injury, you'll have no one to blame but yourself.

FINDING BETTER SOLUTIONS

The Injury Triangle provides a useful perspective not just on what it takes to cause an injury, but also the potential avenues for solution. For someone to get hurt, an object with a sufficient amount of energy must create a hazard, and the presence of a person is necessary for someone to get hurt. These three elements must come together at the same time and in the same place to cause an injury. Thus, removing any one of the three elements will prevent the injury from occurring.

There is the matter of control versus influence. What a supervisor controls is different than what a supervisor influences. *Control* is defined as the ability to determine the outcome. *Influence*, by comparison, is "the ability to produce an outcome without force or direct command"—because the outcome is controlled by someone else. Despite the common wisdom on the subject, for a leader control is far better than influence. With control the leaders get only what they want, and nothing they don't. However, the more likely cause of injury is found in a follower's behavior; a leader doesn't control the behavior of anyone but himself.

When it comes to finding solutions to problems, the application of the difference between control and influence is important to understand correctly. Control guarantees the result: the problem won't return. What does a supervisor control? More than is typically thought: training, qualification, equipment, tools, methods, and procedures. As to the behavior of followers, that is the subject of the leader's *influence*.

Once you understand the correct reasons why something went wrong with an object, the solution usually becomes obvious. In the case of the *Challenger* accident, fixing the O-ring on the solid rocket booster was the easy part of solving the problem. But that didn't mean that once the O-ring was redesigned there wouldn't be any more accidents. The behavior by a big swath of people in the organization also needed to be addressed. That's culture. Changing the way NASA and its contractors operated their business—as the *Columbia* accident later demonstrated—was absolutely needed, and brutally difficult to do.

If you can solve the problem by fixing the object, and there is a simple and easy solution, go for it: write the work order to replace the broken handle on the door, or install a no-skid surface on the walkway to the tool room. Beware the complicated and expensive solution: even if it works, it costs a lot of money. If the proposed solution is time-consuming and expensive, ask yourself, "How else might this problem be solved?" If you can't come up with a better way, ask for ideas. There are always creative thinkers who like coming up with innovative technical solutions. All you have to do is ask them: they'll be happy to help.

But more often, the solution will need to deal with a person, or persons. Before you train the gun sights on the individuals involved, make sure you distinguish normal errors from behavioral choice. People make errors at predictable rates that are much higher than leaders would like them to be. Administering corrective action for what was a normal error is not fair to a good follower, and not likely to have a positive impact on their error rate in the future. Changing the process and system will be required.

As for dealing with the wrong choice of behavior, remember the process to change the behavior of others: by observation, feedback, and the use of positive and negative consequences. Doing that is relatively easy when it's one person's behavior, and brutally different when the behavior is exhibited by everyone in the outfit.

Once you've selected your solution, you need to put it to the test. There are two criteria for evaluating a solution under consideration:

- Effectiveness: How well it will solve the problem.
- Efficiency: How much time, effort, and resources will be required to fully implement it.

The best solutions score high for both criteria. They solve the problem in a cost-effective way. If the solution you've come up with doesn't measure up on both tests, look for other options. They always exist.

CHAPTER 12

MANAGING ACCOUNTABILITY

> We have listened to our stakeholders' concerns that a lack of individual
> accountability undermines the…ability to rebuild trust and to
> move forward.
>
> —*Chairman of Rio Tinto*

In operations dealing with problems is an everyday part of the front-line leader's job. Every once in a while, the problem or event has consequences significant enough to get management's attention, and that's the point where holding people accountable often becomes an important part of the conversation.

10:40 a.m., Managers' Conference Room

Owing to the seriousness of the incident—there had been a huge product release and the potential for an explosion and fire—the general manager had flown in from headquarters to sit in on the investigation review. The meeting, which had begun at 10 a.m., was drawing to a close. The GM, who had sat silently through the meeting, was about to speak.

"Let me see if I heard this thing right. The two maintenance contractors knew exactly what was required to do the job safely before they began disconnecting the line. So did the leadman assigned to watch this job. Those requirements were documented on the Job Safety Analysis they received—which they signed. But they didn't follow those instructions. When our operator checked on the job, he found them working but not complying. He said nothing. When the technical specialist visited the job and saw the problem, he said nothing. Is that right?"

The members of the plant management team nodded in agreement.

"So you're telling me five people knew something was being done wrong and not one of them said anything or did anything to correct the problem before the line started leaking. Is that right?"

There was an embarrassed silence in the room. The GM had correctly summed up the incident in brutally frank terms. Finally the contract administrator spoke up: "That's exactly correct, sir. There's really no excuse for what happened."

The GM replied, "This could have been an absolute disaster. People have to be accountable for what happened."

Nobody disagreed with the GM's conclusion. And everyone understood exactly what he meant by people being "accountable."

They just hoped that didn't include them.

Alive and Well at the End of the Day: The Supervisor's Guide to Managing Safety in Operations, Second Edition. Paul D. Balmert.
© 2023 John Wiley & Sons, Inc. Published 2023 by John Wiley & Sons, Inc.

If you took a survey of industrial leaders on the meaning of accountability, the vast majority would agree with the statement, "being accountable is just a polite way of describing punishment. When somebody does something wrong, there has to be a price to pay."

When the global miner Rio Tinto began to execute its plan to expand its iron ore mine in Western Australia, inadvertently included in the expansion area was the most significant archaeological site on the continent. The property Rio mines was leased from the indigenous people and has literally thousands of historically significant sites to be preserved and protected. At the top of the list were two rock-shelters inhabited 46,000 years ago, a time preceding the Ice Age.

These rock-shelters were destroyed by the miner. The property owners, the public and the shareholders were all outraged. One headline read "Rio Tinto Just Blasted Away An Ancient Aboriginal Site". The company's board commissioned an investigation: "We are determined to learn the lessons from this event so that the destruction of heritage sites of exceptional archaeological and cultural significance… never occurs again."

The investigation found mistakes; those were called "lessons." There were numerous ones made by middle managers at different levels and in different functions over more than a decade. The report summarized them:

- Risk management standards were not followed.
- Heritage considerations were not given the priority that they should have.
- Plans were not fully and completely shared with the landowners.
- New research findings were not recognized, shared, and dealt with.
- Different parts of the company had access to different information.
- Silos within the company prevented issues from being properly managed across corporate functions.
- Removal of information from the information management system created a blind spot.
- Significant issues were not escalated to management.

This is a familiar list of issues every large organization struggles with: compliance, communication, priorities, silos, escalation, and culture. They also come straight off the list of the toughest safety challenges every leader faces individually, and that leaders in organizations face collectively. Fair to say the leaders at Rio Tinto did not manage them well, and the consequences were devastating.

As to managing accountability, the Board's view was that reducing the compensation of the current top executives was sufficient. The public wasn't buying. In the view of the Australian Parliament, Rio Tinto "knew the value of what they were destroying but blew it up anyway." The company did not have a rebuttal to the statement.

That required the more drastic action to be taken by the Chairman: "We have listened to our stakeholders' concerns that a lack of individual accountability undermines the…ability to rebuild trust and to move forward." Hence, a later Wall Street Journal headline: "CEO to Step Down Amid Fallout Over Destruction of Ancient

Rock Shelters. Investors had pushed for senior leaders to be held accountable for destruction of caves."

As the CEO put it, "As chairman, I am ultimately accountable for the failings that led to this tragic event." By the time the dust finally settled on the debacle, two more top executives agreed to resign. No doubt the three would agree with the opening statement equating accountability with punishment.

The Rio Tinto case reminds us that, unless it's an act of God, there are always human fingerprints to be found on the causes of failures, big and small. When humans fail, there is always the matter of accountability to be considered. It's no coincidence that organizations that get great safety results manage accountability well. This practice is critical to safety excellence.

But that leaves us with the question as to the leadership practice of managing accountability: is it simply the process of administering consequences? Or is there something else—something different—involved with holding someone accountable?

As a leader responsible for managing safety, you need to know the answer.

ACCOUNTABILITY

The term *accountable* pervades the language: students need to be accountable for their grades; teachers and schools need to be accountable for their students' test performance; political leaders need to be accountable for the way they spend taxpayers' money; top management needs to be accountable when the business fails to make a profit. And people at work need to be accountable for their safety.

It's easy to appreciate the sense of frustration with failure that leads people to talk about accountability. The public was outraged with what Rio Tinto had done to their country's archaeological heritage. Families are frustrated with employers when their loved ones don't come home alive and well at the end of the day. The need for accountability shows up in sports. In the aftermath of a college football game where the referees made not one but two errors in the final minutes of play—wrongfully allowing one team to catch up and win—the conference director of officiating told reporters, "Yes, the officiating crew is accountable. There will be punishment." The victims of the bad call might take some consolation from knowing that, but it didn't change the final score. They still lost because of someone else's error.

Accountability is bound up with problems and failures. It isn't part of the conversation when there's been a smashing success. "Let's have a big parade to hold the hero accountable!" That explains why *accountable* is used interchangeably with *punish*. Most of the public chatter about accountability is really about the lack of consequences: schools produce students who can't pass the tests, but the teachers and principal still get their annual performance increases. Elected officials make bad laws or don't enforce the laws that are written, but still win reelection. Professional athletes with guaranteed contracts don't perform, but they can't be cut or placed on the bench.

With Rio Tinto, the frustration was with management. It is management's job to make sure standards are followed, priority is given to what's defined as the priority, plans are communicated with stakeholders, important new information gets communicated, those with a need for information get what they need, significant issues get escalated and senior management knows what's going on. None of that happened, but a catastrophe did. Stakeholders demanded accountability. Who could blame them?

"Lack of accountability" is the term used to describe the problem, but in the terms of this book, it's simply a lack of consequences. Nothing happens to those who don't perform. In fact, this lack of performance seems to be rewarded rather than punished: the student cuts class, doesn't study, but graduates. The teacher gives a half-hearted effort, but gets a salary increase. The elected official avoids voting on a tough bill, goes on a junket paid for by the taxpayers, and gets reelected. Industrial executives fail, but keep their jobs. No wonder people are frustrated.

But just because the public wants to equate accountability with consequences doesn't mean you have to follow suit. In the practice of leadership, there is a very significant distinction between "holding someone accountable" and "administering consequences." When understood properly, the difference creates one of the most valuable practices for managing safety performance successfully: managing accountability.

DEFINING ACCOUNTABLE

To understand that difference, let's start with another familiar term that needs to be part of the conversation: *responsible*. A good dictionary will provide a long list of definitions for responsible, one of which is accountable.

But in the matter of failure—when something goes wrong—that's not the most useful way to define either term. When something happens, the more appropriate definition of *responsible* is "the cause" or "the agent." If problems have human causes, then those who caused them are the ones are responsible.

The interest in holding someone accountable arises when something goes wrong: the consequences—damage or harm—create the need to hold someone accountable; the worse the consequences, the greater that need. As to who needs to be held *accountable* for those *consequences,* it's the person *responsible* for causing the problem.

Leaving accountable out of this for the moment, when something goes wrong someone was responsible for what went wrong, and there are consequences to be found from what went wrong. Looked at that way, the person responsible was the *cause*, and consequences the *effect*.

Used in that way, responsible and consequences are simply another way to describe cause and effect. Things happen for a reason. When things happen—for worse or for better—the effects are produced by causes.

So, where does that leave accountability? Does accountability even matter?

On that note, we must return to the preceding chapter, Understanding What Went Wrong. As a leader, you can't hold someone accountable if you don't know about the problem; you can't hold someone accountable when you know there was a problem but have no idea who was responsible for its cause.

Thus, the precondition to holding a follower accountable are that the leader (a) knows about the problem, (b) its consequences, (c) who caused the problem and (d) how they caused the problem. As you recall, when looking into problems, consequences come in two types: actual and potential. When someone working from a scaffold drops a pipe wrench and it hits the ground, the actual consequence—damage—is limited to the wrench; if it hits someone working below, the potential consequences include a fatality.

An event creates consequences, and an event may cause a leader to *administer* consequences to the person responsible for the event in the form of discipline, corrective action, even termination. But calling the administration of those kinds of consequences "holding someone accountable" gives away a powerful leadership and influence practice. Administering corrective action and holding a follower accountable must be viewed as separate leadership practices.

That begins to set up a different meaning for the word accountable: "answerable; obligated to provide an accounting or explanation." It's a definition with biblical roots: on the day of reckoning, an account is owed for how one lived life. By this definition, asking questions and giving answers becomes the basis for the practice of holding someone accountable.

Looked at in this light, being *held accountable* takes on a much different meaning. It's a discussion with a follower led by a leader—note the use of "held" — taking place in the aftermath of a failure. Not a lecture or a Stump Speech, it's a two-way conversation with an important purpose (see Figure 12.1):

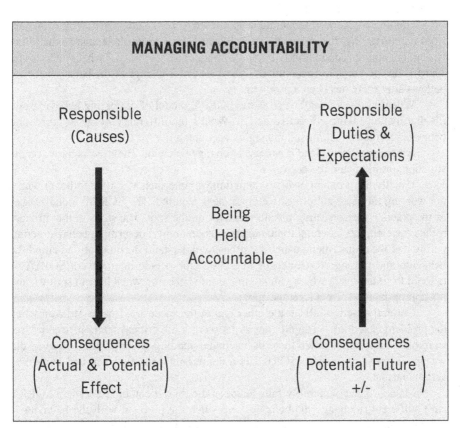

Figure 12.1 In Managing Accountability, a leader will draw into the discussion the two different meanings of "responsible" and "consequences."

- Establishing ownership for what was done wrong
- Appreciating the consequences from what was done—actual and potential
- Re-affirming the duties and responsibilities of the follower and the expectations the leader has for the follower
- Setting the potential future consequences should behavior not change for the better

These four points combine to form the framework for holding a follower accountable. They're based on two different meanings for the words responsible and consequences:

- Responsible means both "cause" and "duties and expectations"
- Consequences mean both "effect" and "potential future corrective action"

The framework for holding a follower accountable is the same no matter what the problem. But, when safety is the problem there's one important difference that must be kept in mind: put in play are the most important consequences possible. It is the Case for Safety: people's lives, the impact on a family's ability to earn a livelihood, the effect on friends, coworkers, and even supervisors. Yes, there might be corrective action, too, but that pales in comparison to The Case.

Holding a follower accountable is a management duty, separate and distinct from administering discipline. It is not the investigation to understand what went wrong: managing accountability begins with an understanding of what went wrong. The practice of managing accountability is intended to change future behavior for the better—after there has been a problem.

When it comes to safety, so is the SORRY model of correcting behavior, and for that matter, taking corrective action. What's the difference between correcting behavior, corrective action and managing accountability?

All are practices a leader can use to change behavior. Their use is based on the situation, problem and the person.

Usually the problem involves something simple such as a good follower who's not wearing the personal protective equipment required. The SORRY model serves as the practice for correcting unsafe behavior on the spot. The steps in that process reflect the elements essential to managing accountability: describing behavior, comparing it to the requirement, using questions to understand the motivation behind the behavior, and putting consequences in front of the person or involved. SORRY is tailored to situations in which you see the unsafe behavior, when it's not serious, and it's appropriate to correct it on the spot.

Corrective action falls at the other end of the spectrum. It is undertaken when the problem is serious—a significant violation of a life critical safety procedure, for example—and comes after there is an understanding of what the follower did wrong. In most operations, SORRY is a commonplace occurrence and corrective action unusual.

Managing accountability falls between the two. It can be used when SORRY isn't sufficient to change the behavior and when the problem with the behavior is more serious, but doesn't rise to the level where formal corrective action is called for.

Managing accountability is not reserved for use with poor performers; there are situations where the best and the brightest benefit significantly from being held accountable by their leader. The supervisor who found two followers doing a cleanup job in the yard and not wearing their PPE would do well to hold his senior follower accountable for the poor example set for the new team member.

But accomplishing that requires the leader to know *how* to hold a follower accountable.

MANAGING ACCOUNTABILITY

Managing accountability takes place in a face-to-face conversation between the leader and the follower. Bear in mind that accountability is a two-way discussion but with a specific purpose: to improve performance. Properly executed it requires the leader to put questions front and center—and listen to the answers from the follower. It is a conversation, but not a pleasant one. There is a time and place for those questions: that time isn't in the middle of a crisis, and the place is not in front of the rest of the world—or right in the middle of an accident investigation.

You might think otherwise. After all, an investigation is a search for the truth about what went wrong, and a good investigation is based on those fundamental questions: who, what, when, where, why, and how. If holding someone accountable is about asking questions, aren't those the perfect questions to ask in the process?

Not necessarily. Holding someone accountable and understanding what went wrong both involve asking questions about failure. Both involve understanding the cause of the problem and recognizing and correcting the consequences. But the typical investigation stops when all the facts have been discovered.

That might be the point for corrective action. Sometimes it is. Holding a follower accountable digs deeper into the process of understanding the underlying human behavior involved. The questions are tough, soul-searching, and the answers are expected to be honest. In a sense, it is a thoughtful examination of conscience, deliberately led by the leader. But the point of it is to change future performance.

In our everyday use of language, "What were you thinking?" has evolved from a question to an insult: a saying that makes a point, with no intent to listen to any possible explanation. That's exactly opposite of the process of holding someone accountable: the salient questions are indeed, "What were you thinking?", "What have you learned from this episode?", and "What are you going to do in the future?"

It's exactly those kinds of questions that are at the core of holding someone accountable. An explanation is owed, not to find out what happened, but to examine the reasons for that behavior and to ascertain the follower owns their behavior. There must be the recognition of the consequences, real and potential—and of the actions that will ensure change in the future. Asking those questions and getting them answered well is the process of "holding someone accountable."

Given those questions and objectives, expecting complete candor in answering those questions in front of an accident investigation panel is asking too much, and may be extracting too great a punishment—particularly from a well-intentioned employee. Those questions are better asked in the privacy of the office, after the investigation is complete.

On rare occasion, the management of accountability as it relates to safety plays out on a public stage. One was the extraordinary three-day public hearing held on the 2006 explosion at the Sago coal mine. That tragedy claimed the lives of 12 miners, 11 of whom perished while waiting for rescue crews. As the events unfolded, the news coverage was riveting: a 40-hour vigil while rescue efforts were undertaken; the rescue of one surviving miner; a false report of the successful rescue of the others; and finally, 45 minutes later, the truth of the discovery of those others, who had been fatally injured.

At the public hearing, family members showed up with questions. For the first time in a mining investigation, family members were given the opportunity to ask those questions of mine safety officials, the investigation team, those involved with the rescue, and, yes, those who mismanaged the incident information. They wanted to know if the accident could have been prevented, why the supervisor's pre-shift safety inspection report hadn't been reviewed by the oncoming shift, and why the rescue breathing equipment hadn't worked. They asked the mine manager, "Why didn't the mine have a rescue team?" By all accounts it was an emotional session, with tough questions and plenty of criticism leveled at everyone.

Except for two people.

The two rescue workers who were responsible for the false report of a success-ful rescue made their report to the families. "We apologize for any problems or heart-ache our miscommunication caused. That was not meant to be." They were applauded, and many family members moved to the stage to hug the rescuers.

The Sago hearing showcased the process of accountability in a very public and emotional way: in the aftermath of a tragedy people asked the questions that needed to be asked and got them answered honestly. It was as tough a conversation as can be imagined, no doubt causing those in management to own their behavior and fully appreciate its personal cost to family and friends. This was far more than get-ting answers.

It doesn't require a tragedy or a public stage to hold a follower accountable. Just like SORRY, there's a simple process to follow that takes five steps. Call it the **Five Ss** for managing accountability.

Holding a follower accountable is best done across the desk in the privacy of your office. It's never a good practice to criticize followers in front of their peers, and you want this to be a two-way conversation between the two. You're looking to create honest introspection, so be prepared to give your follower the time to think and speak and not be interrupted.

Let's use the dropped pipe wrench as the event, but alter the facts as to the consequences: instead of falling to the ground injuring no one, the wrench first struck and damaged operating equipment and then bounced off the equipment, striking and injuring one of those working below. This was a serious event; hence the intent to hold the pipefitter accountable.

1. **Situation:** Begin the conversation with the situation. What was the problem and how did you as the leader become involved?

Remember, this is not your investigation into the problem. Holding a follower accountable is done based on your understanding of what went wrong and it's very likely the matter has already been discussed with the follower: you've already heard what they had to say. That you do not want or need to replay.

But you do need an icebreaker and you do need to set the tone and purpose of this meeting.

"As you know, we have completed our investigation into the pipe wrench you dropped last week while working on the tank repair. When it was dropped, it first struck the pump and then bounced off and hit and injured a subcontractor who was working below. I want to continue the discussion of the event."

By saying "continue the discussion" you have communicated the topic of discussion and signaled your follower hasn't been summoned to your office to receive a letter of reprimand.

The follower might be tempted to reply, "Boss, I already told you what happened in the investigation. I didn't mean to; it was an accident."

This is the benefit of having a step-by-step process to follow. Time to move immediately to the next step in the 5 Ss.

2. **Significance:** Explain the significance of the situation, why is it worth your time to examine and fully understand what happened—and why it happened.

Explaining the "significance of the situation" is just another way of saying "time to talk about the consequences of what went wrong." Here's where your understanding of the consequences actual and potential is vital. Consequences are what get the follower's attention and cause the follower to take the situation seriously. Working three levels up, dropping a pipe wrench did cause a lot of damage to both equipment and person, and it could have been a lot worse!

"Looking into the event, as I did last week, the damage to the pump was significant. It will have to be replaced. And as you know, when it bounced off the pump, it hit the contractor who was working next to it. He was very fortunate that struck him across the upper arm and only left a bruise."

Those were the actual damages; you now describe the potential damages: what could have happened.

"When the pump was hit, it was running and the pipe it was connected to might have been broken, causing a major leak of the contents in the tank. You do know it what's stored there is highly flammable. As to the contractor, he easily could have been killed. A few years ago at a construction site when someone dropped a tape measure it bounced off a crane and killed an innocent bystander. I am not a believer in good luck, but we had our share on this one. This could have been a disaster."

No need to embellish, but you should think through all of the consequences, actual and potential and be prepared to describe them.

3. **Specifics:** What you know the facts to be in the situation. Facts are what can be shown to be true.

The follower may know all of the facts, some of the facts, or precious few facts. It's not necessary to recount them all. Failures usually come with many fingerprints; you're not having this meeting to focus on the shortcomings of others. You'll need to determine what information is necessary to share to focus the conversation where it needs to be.

"When I looked into the event, the big issue was with the failure to tie down the wrench. As you know, it is a requirement that all hand tools used in situations like this have to be tied. Your wrench was not. The purpose of having the procedure is to prevent situations like this from happening."

Those facts are more than sufficient to focus the conversation where it needs to be: changing future behavior in the direction of compliance.

4. **The other side of the story:** What is the story of the person or people involved? What else do they know? How do they see the situation? That will lead to the more important questions that hold people accountable: What were they thinking? What have they learned?

The essence of managing accountability is found in this step of the process. You want the follower to own their behavior, appreciate the problem they have caused, its potential consequences, look inward and recognize their decision-making process that led to their behavior. That is a lot to ask for, and with some followers, achieving that may be impossible. But for many, this will be a Moment of High Influence of the best kind.

While the first three things you have said lay the groundwork, managing accountability is intended to be a discussion: a two-way conversation, but one where you're asking the questions and using them to lead your follower in the direction you want. You're asking Darn Good Questions, albeit in what is a tough conversation. This will not be easy on your follower, nor should it be.

That said, you do want to get the follower talking. So best to start off with asking a question that will be easy for them to answer.

"So, what's your side of the story? Why wasn't the pipe wrench tied off?"

If you've looked into the event and understand what went wrong, you already know what the answer will be. But it doesn't hurt to start there, and then move to the tougher to answer questions. Those you will have to determine on a case-by-case basis, and you should come to the meeting with a prepared list of questions you plan on asking. Your questions and your follower's answers are the core of the process.

As to what else you might ask, here are few questions that usually serve the purpose of holding someone accountable:

- "What have you learned from this experience?"
- "If you had it to do over, what would you do differently?"
- "Why did you think doing it that way was an ok thing to do?"
- "How would you feel if what you did caused someone to be seriously injured?"
- "If this had turned our badly, how would you explain what you did to... (take your pick: your family, the victim's family, the press...)?"

Managing accountability should lead to the discussion of the performance expectations you have for those you supervise. They may be written—policy and procedure—or they may be what you've told those you supervise— "You always need to get confirmation that the people working above them. Expectations are critical to managing accountability well. You can't hold people to a standard they don't know about. If there's a problem and they're hearing your expectations for the first time, the failure is yours as the leader.

That makes another point in managing accountability: in asking the questions that need to be asked, you might learn something you didn't know—and something that might well be key to preventing the problem from happening again.

5. **Steps:** What needs to be done to prevent this from happening again?

The point of exercise is to change future behavior so that something like this doesn't happen again. That may be as simple as getting a commitment from the follower to do better. The circumstances will dictate the appropriate remedy. Just remember: this is first and foremost the follower's problem. As the leader, you're there to help them do their job well, but you can't do it for them. It's always better if the next steps are their idea, not yours. The follower needs to own them.

"Let's figure out what it takes to make sure this doesn't happen again. I'm sure you don't."

APPLYING THE FIVE Ss

Remember that little dustup over the cleanup job going on out behind the shop? That case was first described in Chapter 5, "Managing by Walking Around." The supervisor, adroitly handled that situation by following the five steps of the SORRY model described in Chapter 8, "Behavior, Consequences—and Attitude!" There was an interesting detail in the case that you may have observed and thought about how to handle: the fact that "the senior—and very best—crew member" was in all likelihood misleading the new team member assigned to work with him. If true, does that problem demand more than simple correction of the behavior of both followers on the spot?

Absolutely. Dealing with that kind of problem is a perfect illustration of the practice of holding people accountable. This the supervisor did the next morning, during a one-on-one meeting in his office.

Here's how the conversation went with Charlie, that "senior—and very best— crew member."

The supervisor has skillfully follows the process of holding someone accountable: first describing the situation, expectations, the potential seriousness of what happened, and what the leader knows about the specifics, then asks the first question that begins to get to the heart of holding someone accountable: *What's your side of the story?* From the answers to that question and the ones that followed, the two crafted a plan of action to make sure the problem didn't resurface. At the end of the conversation, the supervisor asked Charlie for a commitment to always set a good example.

8:05 a.m., Supervisor's Office

The supervisor motioned for Charlie to close the door and take a seat. Charlie suspected what the topic of this private conversation was likely to be when the supervisor asked him to meet up in the office after the morning's safety meeting.

"Charlie, as you know, yesterday when I checked on the cleanup job, both you and Ron put your safety gear on only after you saw me. When we talked about the problem at the time, I explained to both of you what I've seen happen when even the basic safety rules aren't followed: people can get seriously hurt, and nobody wants that to happen.

"But there was a second problem going on and I think you know what that was: as the senior guy on the job, you have an obligation to set a good example for the new guy, and to help make sure he follows the rules and learns how to work safely. That's what I would have expected from someone like you.

"But it didn't happen, and the consequences could have been significant. This really is a serious problem. I only know what I saw yesterday: work being done, but without compliance with the safety rules. I haven't heard your side of the story—at least not on this part of the problem. What can you tell me about what happened and why it happened that way?"

Knowing how to manage accountability—starting with what holding a follower accountable really means—gives a leader a powerful tool to manage safety performance. So does knowing the right questions to ask to hold someone accountable—and how to ask those questions in the right way.

In the next five minutes of his discussion Charlie was asked the following questions:

"When you first started working here, what did the senior workers you looked up to do in situations like this?"

"How would you have felt if the new team member you were working with had gotten seriously hurt?"

"What have you learned from the experience?"

The supervisor had one more advantage going in the process of holding a follower accountable: a relationship that provides not only credibility, but also the likelihood of honest replies in answer to the questions. Including the answer that "Boss, everyone in the shop knows you're a creature of habit. We all know exactly when to expect to see you out on the floor. Frankly, you showing up in the yard came as a huge surprise."

Finally, in holding Charlie accountable, the supervisor learned something else important about their practice of leadership. Knowing the truth always plays to the benefit of the leader.

CHAPTER 13

MANAGING SAFETY SUGGESTIONS

> I'm always on the lookout for ideas. Most of us don't invent ideas.
> We take the best ideas from someone else.
>
> —*Sam Walton*

Whether it's a breakthrough idea or what seems like just another complaint, every safety suggestion is a Moment of High Influence. Anyone making a suggestion is engaged, thinking about a problem or a solution, and in a high state of readiness to be influenced.

True, but simply because the suggestion has something to do with safety doesn't mean it's always a good idea, or that a busy leader should feel obligated to immediately drop everything else and devote their undivided attention to managing the suggestion.

Case in point.

6:30 a.m., Friday Morning, In the Office

Friday is usually the day of the week when everything goes haywire, and this Friday was proving to be no exception. The supervisor's day started well ahead of the start of the shift: in the office, preparing for the 7 a.m. toolbox safety meeting. The task was constantly interrupted by the phone: overnight, the line had to be shut down, and getting production re-started was to be the day's top priority.

The supervisor promised to meet the engineer and the planner out on the job to sign the permits allowing the repairs to be started—as soon as the safety meeting was finished. No production problem was going to interfere with getting the crew properly aligned.

However, the supervisor was determined to see to it the safety meeting would be short and sweet.

Alive and Well at the End of the Day: The Supervisor's Guide to Managing Safety in Operations, Second Edition. Paul D. Balmert.

7:10 a.m. Safety Meeting Ending

As the meeting was ending, the supervisor was putting on the hard hat and heading out the door. Standing in the doorway was the newest member of the crew, Ron. He had a big smile on his face, and began on a positive note, "Boss, I know you're really busy this morning, but I got an idea for you."

Knowing the line was down and people were waiting, the supervisor replied, "Ron, I'm really busy. Can you save it for later?"

Ron didn't seem to understand. "It'll take just a minute. With the time change, we're now starting the day in the dark. Where you've got me assigned, I'll be right next to the road. With the construction project, there's a lot of inbound traffic and equipment coming in and out. I'm worried that they might not see me."

If the supervisor was thinking this was a case of taking one of those "best ideas from someone else" they were in for a disappointment.

Ron continued, "So what do you think about big wide blue reflective racing stripes down both sides of my hard hat? That way everyone will be able to see me in the dark. Plus, it will make my hard hat look really cool!"

The supervisor couldn't help but wonder what Ron had learned in his new hire safety orientation: company policy requires that nothing can be affixed to the outside of a hard hat.

A Moment of High Influence? Absolutely! A good idea? An entirely different matter. What's the best way for the leader to manage a safety suggestion like this?

THE FIRST THREE QUESTIONS

No matter how the safety suggestion is received—by face-to-face conversation, as was the case here, getting it over the phone, by e-mail, or reading one submitted as part of a formal safety suggestion program—there are three questions a leader must always first ask in the moment:

1. How urgent is the problem?
2. What should I say to the person?
3. What should I do next?

Consider those as the First Three Questions. Asking them, and in that order, provides the leader with a tool to successfully manage any safety suggestion. Here's why each question matters, and the answers in real time.

Question 1: How urgent is the problem?

The first order of business is urgency or priority: how much time does the leader have to deal with the suggestion. Just because the suggestion involves safety does not mean it must become the top priority. Every leader has a plateful of activities competing for their attention; time needs to be allocated wisely. Doing that becomes easy and obvious—once the definition of a suggestion is understood.

A suggestion is the combination of problem and solution. (see Figure 13.1). To give this definition an additional level of precision, the situation may be not so much

SAFETY SUGGESTION

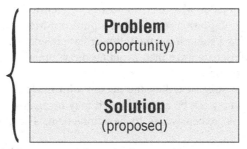

Figure 13.1   By definition every safety suggestion contains two parts: problem and solution. The nature and severity of the problem should always determine urgency or priority.

a problem as an opportunity for improvement, and a proposed solution may not be the best way to deal with the problem.

This is a definition markedly different from what leaders typically think: a suggestion is an idea about a solution. Problems brought to the attention of the leader are suggestions, too. Looked at that way, leaders get a lot more suggestions than they think. They just don't always come with the "good idea" of a proposed solution.

There can be a report about a problem, which might sound like a complaint: "Every time it rains, the stairs get really slick." There may only be a solution: "I think we should put non-skid material on all the stairways."

When the other part of the suggestion is missing, one obvious step to take is to ask for it. If someone raises a problem, ask, "What do you suggest we do to solve the problem?" If there is only a solution, ask, "What's the problem this is designed to solve?"

In the scenario involving the suggestion made at the end of the Friday morning toolbox meeting, the supervisor was fortunate to hear about both the problem and solution: reflective racing stripes was proposed as the solution to the problem of poor visibility early in the morning. That came from the newest member of the crew, who deserves to be given a great deal of positive reinforcement, even if putting tape on the outside of a hardhat is against the rules.

As to the all-important matter of priority—the response time to be given to the suggestion—that is a function of the nature and severity of the problem. Therefore, the first step is always to understand the problem. What is the problem? Where is it? When will it show up? How bad will it be if it does? If in doubt, ask those Fundamental Questions and listen to the answers. They will set the proper priority.

The problem defines the priority of the suggestion. A serious and imminent hazard that demands immediate attention. A brilliant solution for a problem that is guaranteed not to show up for a year need not be acted on right now.

That may be simple common sense, but in the heat of battle it can be easy to overlook a serious problem because the proposed solution is totally unworkable or against the rules. The best way to avoid falling into that trap is to simply sidestep any proposed solution and first focus on understanding the problem.

Question 2: What should I say to the person?

In the practice of leadership, the first words the follower hears from the leader count for a lot. No matter what the problem or the feasibility of the solution proposed, every safety suggestion represents a significant Moment of High Influence. Remember, when it's a suggestion, the follower has created the Moment. The leader needs to immediately recognize that, avoiding the temptation to pass judgment on either problem or solution.

That does not suggest telling the person who made the suggestion, "Good idea." The solution may not be a good idea: it may be against the rules; it may not work; it could be so expensive as to be impractical. There might not even be a problem.

The best practice is to simply say, "Thank you for thinking about safety" or "…getting involved" or "…bringing up a problem" or "…offering a solution." Saying that clearly conveys the leader recognizes a follower has taken the time and effort to make a suggestion. That the problem might be serious or trivial or the solution brilliant or unworkable really doesn't matter to the Moment. For safety suggestions, it's the thought that counts.

Question 3: What should I do next?

"Do next" does not mean determining the final resolution of the suggestion; it simply defines the next step. One of the many good things about managing safety suggestions is the wide range of potential next steps that can be considered. Each option has built-in advantages; most have limitations; seldom is the option to either agree with the proposed solution or do nothing.

- **Fix the problem:** Actions speak louder than words. Whether by means of the proposed solution, or by counter-proposing an alternative approach, there is nothing better than actually fixing something. Seeing evidence that a problem has been fixed is likely to prompt others to bring safety problems and solutions forward. That can't be anything but good.

  One downside to just fixing the problem is that a great solution can go unnoticed by the rest of the operation. When someone working for you comes up with a great solution, be sure to share it.

- **Enlist the Safety Department:** This is a form of delegation, a staple of every leader. The safety staff is likely to have ideas as to alternative solutions and may even have the resources to solve the problem for you.

- **Submit a safety suggestion:** Formal safety suggestion systems are designed to provide an effective means of evaluating and implementing solutions. These systems direct problems and solutions to independent evaluators, and when a new idea is suggested, a formal suggestion system can migrate the solution across the organization.

  There is this downside: many have a bad experience with a formal suggestion system. They can be slow and cumbersome; one bad experience can be enough to turn off a follower permanently.

- **Take it to the safety committee:** Like a safety suggestion system, a safety committee may serve the function of independent evaluation. A safety committee can also facilitate the migration of a new idea across the organization.

But, like a suggestion system, committees can be places where good ideas go to die. If you send the suggestion to a committee, be sure to keep track of it and provide regular feedback to the follower who made the suggestion.

- **Take it to your team:** Taking problems and proposed solutions to the rest of the team can be a great way to build support, improve the solution, and begin implementation. The group may have other information about the nature of the problem ("It's happening all the time"). They may also come up with alternative solutions and improve upon a solution.

  Caution: peers are very capable of putting down someone else's idea. If you decide to take a suggestion to your team, be sure to begin by recognizing what you like about the suggestion, and impose some rules for discussion, such as those for brainstorming, described later in this chapter.

- **Designate the person making the suggestion as "champion."**

  The follower making the suggestion may be a subject matter expert and perfectly capable of selling their solution. On the other hand, designating the follower as the "champion of his or her idea" and sending them out to do battle with the rest of the organization represents dereliction of duty. Case in point: asking a new hire to advocate their solution. It can prove to be such a negative experience for the follower that they will think twice before submitting another suggestion.

- **MBWA:** Going out to the scene to see the problem firsthand is a perfect application of Managing by Walking Around. It's the one next step that never has a downside, other than requiring an investment of the time and attention of the leader. If you want to make a friend for the rest of your life, take the follower who made the suggestion with you. What you see while you're there will tell you what you need to know about the problem and the proposed solution.

- **Stopping the Job:** The perfect application of the principle of control would be to simply "stop the job." Taking that step guarantees the outcome—the problem causes no harm—and it provides time to come up with an appropriate solution.

Back to where the chapter started: Ron's safety suggestion, made to his supervisor in the middle of a production crisis. Understanding the First Three Questions necessary to manage a safety suggestion, his supervisor executed each perfectly.

Here's the rest of the story of what happened at 7:10 a.m.:

After a moment's reflection, the supervisor replied, "Ron, thanks for thinking about safety. I really appreciate that this early in your career you've got your head in the game and are recognizing potential hazards that might get you hurt.

Let's do this: it should be light out there by 8 a.m.. Why don't you stay here in the office area until then? You can use the time to read up on the safety procedures you need to know as part of your training program.

I've got a meeting to go to now. I'll be back by eight, and we'll go out there together, survey the situation, and decide what we need to do to make sure you're safe."

Ron smiled. "Sure thing, Boss. Thanks for listening."

It was now 7:12 a.m. The conversation took two minutes.

There are times when the practice of safety leadership is that simple.

EIGHT RULES FOR SUGGESTIONS

In the moment of hearing or reading about a problem or solution, the best practice for managing a safety suggestion starts with answering those First Three Questions: How urgent is the problem? What do I say? What do I do next? From those questions, a set of rules for managing safety suggestions can be derived.

Rule One: It's the nature and severity of the problem (not the feasibility of the proposed solution) that determines the priority of the suggestion. In the heat of battle, be sure to focus on the problem first.

Rule Two: Always thank anyone who turns in a suggestion. Saying thanks costs nothing, and says nothing about either the nature of the problem or the feasibility of the proposed solution. But saying thanks is the best way to recognize this Moment of High Influence.

Rule Three: Give feedback promptly. When you hear or read a safety suggestion, acknowledge it. Then reply: tell the follower what you plan to do with the suggestion. If you delegate the problem to someone else to investigate or reply, remember that you still "own" the suggestion and are obligated to see to it that there is a timely reply.

Rule Four: Shoot straight. Don't tell people, "That's a great idea" when it's not. They'll see right through that, and you'll wind up losing credibility. But there is a big difference between shooting straight and shooting from the hip. "If you had stayed awake during the training you would have known that what you're suggesting is against company policy" may be true, but it's also a sure way to send someone away unlikely to make another suggestion.

Rule Five: If the suggestion is unworkable, offer advice on how to improve it. Doing that doesn't have to put you to work redesigning someone else's idea; if the proposed solution will cost a million dollars, take a year's worth of engineering, or violate a standard, explain those problems, describe the criteria for a successful solution, and ask, "How else might this problem be solved?" For every problem there can be multiple solutions.

Don't make the mistake of thinking every idea has to be perfect. Studies of successful innovation have provided some very useful insight into the process of converting a new idea into proven technology. In the early stages, virtually every breakthrough idea had significant flaws and was not workable. But the ideas that have revolutionized technology—from the light bulb to the personal computer—went through an evolution by which problems were eliminated or reduced until the product became useable. At that point the new product still wasn't perfect, but it was more than good enough to buy.

Rule Six: If the suggestion involves something under your control, act on the suggestion or explain why you don't. If a suggestion isn't workable, sloughing it off to someone like the safety department is a waste of the valuable time of the business. People can handle rejection. But they'll accept no as the answer far better if they're told why.

Of course this will expose your logic to the light of day, so make sure your reasoning is sound.

Rule Seven: Communicate face-to-face. Even when the system provides a written response, face-to-face contact has several important advantages. The investment of time in seeking out the person making the suggestion sends an important statement about a leader's commitment to safety—and demonstrates a genuine interest in the follower. It provides the opportunity to listen to what someone has to say about a problem or a solution. Formal suggestion systems are hampered because they don't provide that opportunity: it's entirely possible that the real problem isn't what the person is complaining about. There may be better solutions to the problem than the one proposed as part of the formal suggestion.

Limitations on that face-to-face communication are an operating reality: not every leader has the luxury of daily contact with those supervised. If the problem is minor, scheduling a review of the suggestion during "the next visit to the area" is a perfectly appropriate response. Just be sure to follow through on that commitment.

Rule Eight: Recognize successful suggestions. To encourage safety suggestions operations have taken to using a variety of incentives ranging from belt buckles to gift cards. In principle there isn't anything wrong with using incentives to encourage and recognize desired behavior. But in the case of a safety suggestion, it is unnecessary—and misses the point entirely. People turn in safety suggestions for one of two reasons: they want to see a problem fixed or they want to see their idea implemented!

Understanding what motivates suggestions makes determining the appropriate incentives easy: for those who identify a problem, fixing it is reward enough. For the innovators who want to make the world a better place, implementing *their* new idea is all the reward they need. For either type, publicizing successful suggestions is the surest way to get more. Giving credit for those who made the suggestions never hurts.

GOING ON THE OFFENSIVE

If you think getting more safety suggestions from your followers sounds like a good idea, you don't have to cross your fingers and hope they'll start showing up. You can prime the pump: getting your crew thinking about making suggestions about problems and situations that matter to you. Here's a simple process for doing that:

- Read the accident and audit reports to learn about the kinds of problems being experienced in your operation. Those suggest potential problems in search of innovative solutions.

- Explain those problems in your safety communications—safety meetings, toolbox safety sessions, and informal discussions—telling followers you're in the market for ideas about how to solve them.

- Encourage followers to make suggestions about the types of problems under your control.

- During safety meetings use the brainstorming process to generate ideas about how to solve a safety problem. That process begins with a question, like

"What ideas does anyone have about how to solve the problem with...?" Write every idea down, and don't permit the evaluation of any idea until the brainstorming process officially ends. Once the process of listing has been done, take the best ideas, refine them, and convert the result into safety suggestions.

When it comes to tackling the tough challenge of managing safety performance, you can use all the help you can get. Safety suggestions are a great way to get help, and gain a significant number of side benefits as well.

CHAPTER 14

SAFETY MEETINGS WORTH HAVING

> No sinner is ever saved after the first twenty minutes of a sermon.
>
> —*Mark Twain*

When was the last time you went to a really good safety meeting, the kind you and everyone else in the room found worth their time to attend? Or are your safety meetings more like this one?

7:45 a.m. Friday morning: Monthly Department Safety Meeting

Fifteen minutes into the department's safety meeting, it's the manager's turn to present the "safety topic of the month"—talking points coming straight from HQ. The only consolation: the audience has already mentally deserted the room, the predictable effect of opening the meeting with the dreaded review of the safety statistics—known by all as "Death by PowerPoint."

Not that the leader could blame anyone: she hadn't been paying much attention, either. "Why do we go through with this charade," she asks herself, "month after month, year after year? There's not a person in the room who wouldn't rather be out on the job, doing something productive. Including me." As she thumbs through the material about to be presented, "And what were they thinking when they told us we had to present this in a safety meeting? It's a complete waste of time."

Twenty minutes later, the attendance sheet was passed around. All participants signed, indicating they were present in the meeting.

Safety meetings have a noble purpose: helping followers do their jobs safely. But too often, these weekly, monthly, or quarterly safety gatherings wind up a misuse of valuable time. When it's seen that way, the evidence from the audience is unmistakable. Fighting boredom, they fidget, look at their watches, doodle, work on the crossword puzzles, engage in side conversations, or drift between sleep and wakefulness, counting the minutes before the meeting is over and they can get out with doing something useful, like work or lunch.

Safety meetings didn't start off that way. Imagine the first safety meeting ever held: it probably took place in a steel mill in the 1800s, when the mill manager took the unprecedented step of stopping production and calling everyone together to talk about safety—because it was terrible. Stopping production spoke volumes about the leader's commitment to safety, but that was a very long time ago. More than a century later, their novelty has long since worn off. Today's reality is that many safety meetings are simply an exercise in "ticking the box" providing little of value to the cause of sending everyone home, alive and well at the end of the day.

What a waste of a Moment of High Influence!

You might question how a meeting putting people to sleep could possibly be a Moment of High Influence: at worst, it seems like a missed opportunity, doing no harm to the cause. Consider the message a leader sends to followers by holding a safety meeting like the one described: coming across loud and clear is that the leader doesn't take safety seriously. That the leader is going through the motions is stunningly obvious to every follower in the room. That is a significant Moment, but one unrecognized by the leader.

Managing safety performance is already tough enough; why add to the challenge by doing something like this on weekly basis? The leader would be better off having no safety meeting.

But since the institutional practice is unlikely to be eliminated, it's time to figure out how to hold safety meetings worth having.

THE PROBLEM WITH MEETINGS

Why do many safety meetings fall short of the mark? The problems are obvious. First, much of the content strikes participants as boring and irrelevant. Sometimes because that is exactly what it is. Standardized content to be covered with everyone doesn't always apply to the entire audience. It's often information they've heard before, like refresher training required to be covered even if everyone in the audience has heard it many times before. Even new information—no matter how important—isn't necessarily seen as interesting.

That's content; then there's the matter of delivery, which is the even bigger problem. Presentations are, by definition, a form of one-way communication. At least up to the point where the presenter asks, "Any questions?" Bored participants hope there are none.

A large audience may not be uniformly interested in the topic, and the size of the crowd dampens audience participation. Yes, with enough preparation and practice it is possible to lead a large safety meeting that would be effective: there are professional speakers who do that for a living. But as a practical matter, supervisors and managers are too busy to put that kind of time and effort into building a stimulating, memorable safety meeting. And public speaking is rarely a core competency.

Finally, there's the matter of time—and timing. Holding a meeting at the end of a long day or in the middle of the night isn't the ideal time, but as a practical matter, meetings have to be managed around what the work schedule dictates. Scheduling the safety meeting for an hour creates still another problem: the attention span of the participants. Studies of our human brains have found that the average attention span of an adult is at best roughly 18 minutes. The popular video TED Talks have that time limit for exactly this reason.

But even that's an optimistic number: given irrelevant content and a poor presenter, it doesn't take 18 minutes to lose an audience. And once it's been lost, getting people's brains back in the room is even tougher.

When you consider all the factors in play necessary to putting on an effective safety meeting—content, presentation, audience, timing—it almost seems as if the practice was deliberately designed to fail! A leader would be better off not having a meeting, and putting the time saved into effort that might actually provide some benefit.

Still, safety meetings are too important to be allowed to fail. This may be self-imposed, but holding safety meetings that matter is one more tough safety challenge that a leader serious about safety leadership must take on.

Consider what's to be gained from a good safety meeting. Reviewing the steps that help keep people going home safe, learning from the accidents suffered by others, improving understanding and compliance with safety policies and procedures, unearthing and solving problems, and explaining changes in safety policies, productive discussion of problems, coming up with solutions to problems represent high value use of the time of the leader and followers. The audience in a safety meeting might even find those topics interesting. It's not hard to remember at least one safety meeting well worth attending.

The problem with safety meetings has more to do with *delivery* than it does with content: good content poorly presented still makes for a bad meeting. Most content, well presented, can be interesting. If you're the one responsible for either selecting the content or making the presentation, here's the point where the news gets better. There is a simple process to follow that is guaranteed to produce a good safety meeting.

A safety meeting doesn't have to be a smash success to get the job done: a good meeting is good enough. Learning the method is simple; executing it well is not difficult. It's built on a practice you've learned earlier in the book.

BETTER SAFETY MEETINGS

Think back to the last time you were in a really good safety meeting—one led by a leader like you, not by a professional presenter: what went on? People in the meeting probably did the bulk of the heavy lifting carrying on the conversation. They did that willingly because they were engaged in the topic—undoubtedly because they saw its value and relevance to what they do for a living. The discussion led to a beneficial result: people came away from the meeting having learned something important and useful; were motivated to do something better or different; discovered a problem or identified with a solution.

That's what happens in a good safety meeting. The solution to the challenge of running safety meetings that matter is simply being able to produce that kind of effect on a regular and predictable basis. By the process of reverse engineering, the steps to follow to accomplish that become clear. All that is needed are a good topic and presentation method that doesn't "present."

Preparation for a safety meeting begins with content, the subject matter for the meeting. "If only they would send me better material, I could have a good meeting" is a common lament. Providing leaders with material for the safety meeting might sound like a good idea, but all too often what's provided from the front office or safety department is wide of the mark. Not every topic is relevant and important to everyone.

In finding content that fits your followers, there's an unlimited supply of great material for safety meetings sitting right in front of you. It's in your e-mail, on your desk, in the safety manual, in the monthly safety report posted on the bulletin board, in the news, on the internet. You just need to think differently—and better—about what makes for a good source material, where to look for it, and then how to put it in play in a safety meeting. Content needs to be matched with purpose.

Sharing lessons learned is the purpose behind routing accident investigation reports. Every great safety tragedy, and the many more small events and near-miss reports, represent opportunities to learn from the misfortune of others. Bad enough that someone else got hurt—or might have—but at least others can benefit by learning from what happened. That viewpoint frames a fundamental question in these cases: how do we prevent this from happening to one of us?

Existing policies and procedures—they are numerous—can make excellent safety meeting content. All were written because of some past event. Policies are used with varying frequency. Yes, there is periodic refresher training, but what's the retention of that, particularly if the procedure hasn't been used for a long time? Looked at in that light, any safety procedure begs an important set of questions that can drive a very productive discussion in a safety meeting: What's your understanding of the procedure? When we use the procedure, how well does it work? What are the problems when we use it?

Safety performance reports—the numbers—strike many as boring. Dry numbers are. But when the numbers show up on the sports or business pages and involve things like your favorite team or the stock you own, those numbers are anything but dry! The newspaper and television reporters figured out a long time ago how to make the numbers interesting and important: they don't just give the final score or closing price, they tell the story behind the numbers. That's what gets readers to read.

The same logic applies to the safety numbers. Hours worked without an injury, number of safety suggestions submitted, near misses reported, behavioral safety observations, audit scores are all just numbers. But the numbers always pose important questions: what are the numbers telling us? Why are we having fewer accidents? Why aren't people turning in as many safety suggestions? What can we do to reduce the number of vehicle accidents?

In every case, the content—lessons learned, policies and procedures, safety performance reports—becomes relevant, important and interesting when those subjects are examined by questions like these.

"ASK, DON'T TELL" SAFETY MEETINGS

As how to the method of running a good safety meeting, it's simply a matter of running a safety meeting by asking questions. Darn Good Questions! That's why the method is called **Ask, Don't Tell.** By asking Darn Good Questions that meet the four criteria described in Chapter 9, your questions will do the heavy lifting for you. Your followers will be more than happy to answer them, and whey they're doing the talking, you're not doing the presenting.

The solution to the challenge of safety meetings really is that simple—and that easy. But there is more to Ask, Don't Tell than just firing questions at your audience.

STEP 1: PURPOSE FIRST

What it the purpose of a safety meeting? What is your purpose for holding today's safety meeting? Simple as they are, these questions are seldom asked—or answered— other than to say, "We held our safety meeting this month." You might meet the objective, but meeting over, what do you really have to show for the investment of time and effort—yours and your followers'?

Content is what's to be talked about. Content comes from a wide variety of sources, but whatever the content, you need to put it to the test of purpose. By asking those questions, the beneficial purpose for the topic of your safety meeting becomes clear. Establishing clarity of purpose in very specific terms is the first step on the path to holding safety meetings that matter.

Coming up with a good purpose needn't be complicated. Doing that is as simple as completing the sentence, "My purpose in talking about this subject with my followers is that..." When it comes to your objective, think clear, simple—and specific. The more detailed and specific the better: "So that doesn't happen here" sounds good, but comes up short of the mark: what exactly is the "that"?

Good statements of purpose read like these:

- "Make sure everyone understands the hazards we'll be working around today."
- "After the long weekend, get everyone focused on working safely."
- "Increase reporting near-misses."
- The review of the safety statistics can have a useful purpose: "Delve into our performance numbers to understand what we're doing right, and where we can do better."

Your purpose also provides the means to evaluate the effectiveness of your safety meeting. Meeting over, you can ask yourself, "How well did I do in achieving today's meeting objective?"

Determining the purpose is entirely under the control of the leader running the meeting. If you're the leader, and you don't have a clear purpose in mind, you have no one to blame but yourself for a bad safety meeting.

STEP 2: CONDENSE INFORMATION

Most supervisors and managers are quick to point out they are not particularly good public speakers. So, given the opportunity to lead a safety meeting, what do they do? Talk, talk, talk; tell, tell, tell. Otherwise known as public speaking. Talking is not playing to the strength of the leader. When a leader is speaking the best that can be hoped for is that followers are listening.

Public speaking and keeping audience interest high is a tough act. One-way communication puts followers in a passive mode. The simple solution is to stop talking: for whatever content that needs to be explained, convey the information in the fewest number of words possible. How? Create headlines, the way the newspapers do for every story: "Worker crushed when equipment unexpectedly started."

What about all those important details found in the rest of the story?

They're not nearly as important as you might think. Newspapers figured that out years ago: a good reporter always puts the key part of the story in the first paragraph: it's called "the lede." The rest of those seemingly important details often bog down the story.

Stick to the key points that supports your purpose. If your meeting objective is to re-emphasize the importance of equipment isolation and de-energization, summarize the story this way: "A maintenance contractor working for a competitor was performing a routine inspection of turbine during an outage. The equipment had been shut down, but electrical isolation wasn't done. Someone inadvertently hit the start switch."

In three sentences, there's more than enough information to prime the pump for discussion on that subject. Remember, it's your safety meeting: your focus is on your followers. You and they aren't there to do a root cause investigation into someone else's tragedy. Your purpose is that your followers learn from what happened to others so that the same thing does not happen to them!

STEP 3: MAKE THE CONNECTION

When they don't see any value in the topic, people get bored very quickly. Who can blame them?

Don't waste valuable time on content that is unimportant to your audience. But just because you think something is important doesn't guarantee your followers will, too. You don't want to leave it up to your audience to figure out the relevance and importance of the topic: tell them.

Think of this step as "making the connection" between content and audience. One simple sentence is more than sufficient to accomplish that. "The reason why it is important to talk about this tragedy is because we regularly work on equipment that must be properly locked out to be safe."

Do that, and nobody in your audience can say "this doesn't matter to me."

STEP 4: ASK DARN GOOD QUESTIONS

Content summarized, connection made, the safety meeting is the perfect occasion to let Darn Good Questions do the heavy lifting for you. When you're asking questions your team members will be the ones doing most of the talking; you're no longer in presentation mode. Good questions will cause learning and discovering in the process. Asking Darn Good Questions is the perfect antidote to the boring, one-way communication that characterizes most safety meetings.

When formulating your questions, remember the "directional arrow of influence" created by the use of the key words, who, what, when, where, how, and why. Your choice of the key word will focus the audience on the topic the way you want them to.

Be sure to put the questions to the test of criteria defined in Chapter 9, in particular making sure they are safe to answer—as seen by your followers.

Here are some examples of what Darn Good Questions might look like.

- "What do our procedures say we should do in a situation like this?"
- "Who's ever found themselves in a situation like this?"
- "Why do you think a procedure this important wouldn't be followed?"
- "Where could something like this happen here?"
- "How do we make sure something like this will never happen to us?"

A few good questions—two or three—are all it takes to have a 15-minute discussion on any topic. That's usually the length of a toolbox safety meeting. If the meeting is scheduled to last an hour, choose three or four different topics. That keeps you under the limit of the attention span—for each.

These four steps combine to form an effective **Ask, Don't Tell Safety Meeting**:

Step One: Purpose: be clear about your objective

Step Two: Condense the information to a headline and summary

Step Three: Connection: tell them why the topic matters to them

Step Four: Ask Darn Good Questions

Refer to Figure 14.1.

A MOMENT OF HIGH INFLUENCE

Circling back to the Department Safety Meeting, the manager decided this meeting was the perfect opportunity to try out the Ask, Don't Tell method. The day before she had read a story in the local newspaper about a tragedy that took place at the

"Ask, Don't Tell" Safety Meeting Outline	
Purpose	My followers are better able to recognize hazards and better prepared to handle emergencies properly.
Headline	Worker at the University Research Center died from H2S exposure.
Summary	Sixteen-year-old assigned to clean up sludge in a tank. H2S unexpectedly released. Co-worker rescued him but died in the process.
Connection	We know about H2S and can encounter unexpected hazards in what seems like a routine job.
Question	Where do you think we might encounter a situation like this?
Question	If you saw someone suddenly collapse, how would you handle the situation?
Question	What should we do to ensure something like this never happens here?

Figure 14.1   An example of the outline for an "Ask, Don't Tell" safety meeting.

University Research Center in town; she thought it would be an excellent subject for discussion in the morning safety meeting.

According to the article, two people were assigned to clean sediment left when a sixteen foot deep concrete tank was drained. One was a sixteen-year-old student working part time, working with a senior staff member. The cleanup job seemed routine; no PPE was required or used. The two entered the tank and began spraying the sludge with hose. Unexpectedly, hydrogen sulfide gas was released. Both were overcome; before he died, the senior worker managed to get the sixteen-year-old's head above water. The newspaper article described him "as a hero for saving the life of a sixteen-year-old boy while sacrificing his own."

Planning to follow the Ask, Don't Tell model, the leader chose as her purpose "Impressing on everyone the importance of recognizing potential hazards before starting any work." Next she condensed the story into a headline and summary, identified the connection between a university research facility and her operation, and developed three Darn Good Questions.

Here's the way things played out on Friday

7:45 a.m. Friday: Monthly Department Safety Meeting

Fifteen minutes into the department's safety meeting, it's the manager's turn. She began by explaining, "Today, instead of reading the safety bulletin, I want to tell you about a tragedy that happened in town on Wednesday and ask you a question."

The room grew quiet.

"Worker At The University Research Center Died From H_2S Exposure"

Several followers nodded: apparently the leader wasn't the only one who heard about the story.

"A young man was assigned to clean the sludge left in the bottom of a sixteen-foot tank using a hose, working with a senior staff member. They thought this was a routine job, but the water released H_2S. When his co-worker realized what happened, he pulled the young man's head above the water, but died when overcome by the H_2S. The article called him a hero."

The manager went on, making the connection: "We aren't a research center and we don't employ sixteen-year-olds as part time workers. But all know about H_2S and anybody can encounter unexpected hazards in what seems like a routine job."

The body language of the meeting participants made it clear the leader had them thinking.

Next, she asked her first Darn Good Question: "Where in our operation do you think we might encounter a situation like this?"

There was a pause. Her followers weren't prepared for a question. She resisted the temptation to answer her question for them.

Finally, someone volunteered an answer. The manager responded: "That's a great point." That's when the floodgates opened. A succession of opinions were voiced; some better than others, but not one off point. The story hit a nerve.

Five minutes later, her followers stopped talking long enough for her to squeeze in a second Darn Good Question: "What should we do to make sure something like this never happens here?"

8:40 a.m.:

The meeting had ended as scheduled, promptly at 8:30 a.m. But more than half the participants had stayed behind to continue the discussion—among themselves.

Clearly the meeting was a Moment of High Influence, and a good one.

BIG MEETINGS

You might be thinking this model would be great for a toolbox safety meeting. It is. You might also be thinking that it's impossible to run a meeting this way with a big crowd—say, for example, a hundred people in a big meeting hall for the quarterly All Hands Safety Meeting.

Not so.

Picture trying to ask Darn Good Questions to a big crowd. The obvious problem is that few people will speak up, and often those who do are the ones you wish wouldn't. If you look at a big crowd as nothing more than a collection of smaller groups, you can use that to your advantage to manage the process.

In a big meeting room participants are usually sitting at tables or in clusters, often with their friends. After you've set up the subject—headline, summary, connection—simply ask each group to come up with an answer to your first question. If you need to, put the Darn Good Question on a PowerPoint so everyone in the room can see it. That's one case where a PowerPoint slide can actually be helpful.

If the subject is an event that happened somewhere else, your question might be "What should we do to make sure this doesn't happen to one of us?" Ask a good question, turn the groups loose, and it will get pretty noisy. That's a good thing: you now have a room full of engaged participants. Give them a few minutes to discuss the question, and then pull them all back into the same meeting. "What group has come up with a suggestion it would like to share?"

The best way to conduct that kind of debrief is by standing in the middle of the room. Think of it as theater in the round. All you have to do as the leader is to play traffic cop: make sure that one person speaks at a time, and loudly enough to be heard by the rest in the room.

The method works. And when it works, the leader isn't presenting, or even doing the talking. The participants are, and when they do they'll be talking about what you, as the leader, want them to be talking about.

That's a safety meeting worth having!

CHAPTER 15

CREATING THE CULTURE YOU WANT

With the slightest push—in just the right place—the world can be tipped.

—*Malcolm Gladwell*

Every organization has a safety culture. If you want to know what it is, all you have to do is to look.

7:45 a.m., Control Room

It's the first day on the job for a new maintenance technician, but one with many years of experience in the trade. During the safety orientation the day before, the technician was told how important safety was to the company's top management: an injury-free workplace is the goal, and every safety rule must be followed completely. There was a test on the safety rules: the tech passed with a 100% score. Now qualified, it's time to go to work, a member of a crew assigned to perform a field repair.

First stop: the Control Room. The operator in charge of reviewing required safety precautions specified on the Safety Permit before the work starts shows the maintenance crew the document and says, "Just sign here, and you can get to work."

Everyone on the crew signs, except the new technician. Taking the safety orientation to heart, the technician says, rather timidly, "Is that all that's required? Don't you think we should walk out the job and double-check everything like we're supposed to? That's what they told me to do during orientation."

Everybody laughs—except the new maintenance technician.

The operator frowns. "What are you trying to do? Slow the job down?"

The peers on the crew chime in: "Look, you're new here. Nobody ever does that. We always just sign the permit. Otherwise, nothing would ever get done."

After a long pause the newest member of the crew does exactly what everyone else does: signs the form. Why rock the boat? Besides, if that's what everyone else does, it must be what management really wants.

The next day, when the situation comes up again, the new crew member doesn't think twice about signing the permit without walking down the job.

So much for following all the rules—all the time.

Alive and Well at the End of the Day: The Supervisor's Guide to Managing Safety in Operations, Second Edition. Paul D. Balmert.
© 2023 John Wiley & Sons, Inc. Published 2023 by John Wiley & Sons, Inc.

In various forms—in venues all over the world—a scenario like plays out. Management writes the rules. Management tells everyone to follow the rules. But no matter what managers might wish, how the rules are interpreted, and how well the rules are followed, is determined by the followers, not the leaders.

In a word, it's *culture*.

Culture looks subtle, even harmless. So does quicksand. It isn't the least bit difficult to comprehend how the culture works in this brief illustration of how the new maintenance technician finds out the way things really are. It doesn't take the newcomers long not just to understand exactly what the culture is, but more importantly, to conform to it. The first time a situation comes up, the new person learns. The next time, they keep their mouth shut. After a few times through this process, they don't even think twice about it…they know that's the way things are done around here. That's what culture is capable of: undermining the best intentions of the leaders. And once the culture is set in place, it's brutally difficult for a leader to change it.

From the front line to the executive suite, dealing with the culture is one of the great challenges leaders face every day. It leaves leaders frustrated, unable to create the culture they *want* and stuck with living with the culture they *have*.

When it comes to changing the culture, the prognosis isn't good: culture change initiatives are far more likely to fail than to succeed. The inability to successfully change the safety culture has proven to be fatal to the followers working in the culture, and career ending to executives unsuccessful in changing the safety culture.

The former CEO at the oil sands operations giant, Suncor, can attest to the truth in that statement. His was a company that suffered twelve fatalities in less than eight years, making them one of the worst safety performers in their industry. In a letter to the board, the company's high profile activist investor said the safety culture must change. The CEO had been hired to change safety performance—and the safety culture. That didn't happen; instead, a thirteenth fatality did. The CEO resigned the next day. The company's planned presentation to the investment community was postponed indefinitely.

If you're a leader dealing with an unacceptable safety culture you can't just shrug off the problem, leaving it to your successor to solve. The cost in human life and suffering is too great, the impact of poor safety performance on the business too severe, and the performance can be career limiting.

That's the Case for Safety.

DEMYSTIFYING CULTURE

The attention placed on culture is hardly new. In the 1950s, in *The Practice of Management*, Peter Drucker used the term "the spirit of the organization" to describe what today is referred to as culture: "An organized group always has a distinct character. It moulds the behavior and attitudes of newcomers…A major requirement of managing managers is therefore the creation of the right spirit in the organization."

Creating that "right spirit in the organization" is of particular interest to the executive. Not surprisingly, front-line supervisors are more inclined to accept the culture for what it is: something far bigger and pervasive than they can take on. But, with the right spirit in the organization, the work of the leader becomes far easier and the performance can improve dramatically. That explains culture's appeal. But while it's easy to describe that right spirit in broad terms, creating it is an entirely different matter. Frankly, the odds are stacked against every leader: successful culture change is the exception rather than the rule.

Part of the problem is with the attributes used to define the culture. A leader can go on at great length about the kind of "spirit in the organization" they desire: "Our people buy into safety…believe that accidents are preventable…see safety as a core value…safety becomes a value shared by all …people do the right thing because it's the right thing to do." Describing the vision is easy; the hard part is translating these attributes into practice out on the job. Behaving differently is seldom part of the description or the discussion.

Not that there is anything necessarily wrong with those characterizations; they describe what any executive should want in terms of what Drucker called "the spirit of the organization." The problem is those kinds of descriptions drive leaders to focus their attention and energy on efforts that fail to produce a demonstrable change in behavior at the point of execution like "changing belief" and "sharing values." No great surprise, the culture doesn't change for the better and neither does safety performance.

But the problem can be traced back to a point much earlier in the process. It begins with the conspicuous absence of a commonly understood definition of culture on which to base efforts to change the culture. So, we shall begin by demystifying this popular management term of art, *culture*.

Webster's defines culture as "the integrated pattern of human behavior." That definition suggests the nature of the problem, but what has steered many culture change initiatives are that of scholars and the academics. As you would expect, the experts bring a more comprehensive—and complex—scope and definition of the phenomenon.

The noted organization behavior educator and consultant Ed Schein, among his many experiences, worked on the effort to improve safety culture in the nuclear power industry. He suggested that the "term culture should be reserved for the deeper level of basic assumptions and beliefs that are shared by members of an organization, that operate unconsciously, and that define in a basic 'taken-for-granted' fashion an organization's view of itself and its environment." It's fair to say those unconscious, taken-for-granted assumptions drive behavior and the day to day decision-making that define the safety culture, as, for example, a decision in a control room between an operator and maintenance staff to waive safety rules in order to meet the immediate production needs of the business. Hence, the opening scene where the new employee learns about one of those "taken-for-granted" assumptions: independent verification of isolation exists on paper but not in practice.

A real-life example of this view of culture can be found in the Space Shuttle *Columbia* accident. In its investigation NASA pinpointed its culture as one of the root causes, describing culture as "the basic norms, beliefs, and practices that characterize the functioning of a particular institution…the assumptions that employees make as they carry out their work." With Columbia, one of those assumptions was that an unwanted event such as piece of insulation falling off the external tank during launch that did no damage wasn't a problem—it was a "normal abnormality." In NASA terms, it was called "foam shredding." The truth was, falling insulation was a near-miss that was reported but unresolved. Eventually one of those pieces of insulation hit the orbiter and caused fatal damage.

Defining culture as NASA did, at the corrective action step, put leaders in the business of trying to change beliefs and assumptions by their followers. If you're a manager on that mission, what are your odds of success?

NASA's odds weren't all that high. Two decades earlier the investigation report into the *Challenger* accident had blamed information loss among layers of management, as well as the organization's "can-do" attitude, as critical factors in the decision to waive the minimum launch temperature requirements and launch the fated mission on January 28, 1986. This was a case where the leaders did roughly the equivalent of what took place at 7:50 a.m. in a Control Room. The biggest difference: middle managers created and signed a waiver to *their* safety procedure.

The investigation findings caused management to launch a culture-change effort aimed at communication and management decision-making, including giving "time out" cards to its engineers and scientists. If that well-intended effort had succeeded in changing the NASA safety culture, *Columbia* might never have happened.

Another culture expert, James Reason wrote, "Like a state of grace, safety culture is something that is striven for, but never achieved."

If you are a leader in operations managing with a safety culture you're not the least bit happy with, what you really need is some practical advice you feel confident will set you on the right path to create the safety culture *you* want. That's the intent of *Alive and Well*: culture change is not an academic subject with a test graded by a PhD in organization behavior. A poor safety culture can prove fatal to people at work.

So, as a first step to getting on a glide path to real change, let's define culture in a clear and practical way: culture is what most people do, most of the time. That's a definition entirely consistent with Webster's.

This definition aligns perfectly with Liberty County Texas Sheriff Deputy Brad Taylor's take on the subject. The Deputy was called to investigate a workplace fatality, where a truck driver picking up a load of sand was killed at the front gate.

> "It appears that the driver was standing on the side tank of his truck with the truck still in gear and slowly moving when he was knocked off the side of the truck by a nearby post and fell under the wheels of the truck where he sustained fatal injuries.
>
> Deputy Taylor said that sources told him that it is a *common practice* for incoming drivers to open the door of their truck and stand on the running board or outside the gas tank and "swipe" their entry card past the card reader as it is difficult to reach the card reader while sitting in their truck seats."

Using the good Sheriff Deputy's characterization, culture can be summed up in two words: common practice. Yes, there's a case to be made for adding into the mix terms such as climate, assumptions, beliefs, norms, and "what's taken for granted." But all those add-ons only serve to complicate a subject that is already complicated enough.

This simple definition provides both useful and practical insight into what culture is, and begins to suggest where to focus a leader's attention and effort to successfully change culture.

THE IMPLICATIONS

Before getting into the details of what to do and how to do that, it's worth spending time examining the implications of this simple definition of culture.

1. By definition, every organization has a safety culture.

Following Reasons' definition—"something to be striven for but never achieved" —leads to a conclusion that safety culture is unattainable. As a practical matter, every organization has a safety culture: it's simply what most people do, most of the time. Common practice. Most often, it's just not the safety culture the executives want.

2. What most people do describes what the culture is.

In every organization, people do all sorts of different things, differently depending on where they work and what they do: production, maintenance, product distribution, laboratory analysis, accounting, information technology, supervision. The *safety culture* is found in those things people do that is necessary to going home alive and well at the end of the day. In the opening scenario, someone from the Safety Department likely ran the new hire safety orientation; a supervisor qualified and assigned the new technician; there was an author of the Safety Permit; an operator in the Control Room owned the equipment and the permit and authorized the work; the maintenance crew signed off and performed the repair.

Let's name each of these different things a **behavioral dimension for safety**: a specific activity that in some way impacts safety performance. For each activity, you can further drill down to identify many more behavioral dimensions that in some way involve or impact safety. For example, performing the repair can be broken into dimensions, much the same way as a behavioral safety observation would organize them, such as wearing PPE, using the proper tool, following the prescribed method, proper body positioning.

The collective behavior of everyone in the organization—all of the people and all of their behavioral dimensions that in some way relates to safety—would look like a standard distribution curve. Not everyone does the same thing, nor does everyone do the same thing in the same way. The culture is found in the middle of this distribution: what most people do, most of the time. Collective behaviors may be tightly clustered or widely distributed; there will be those at the extremes (see Figure 15.1).

That is not an unimportant point: the culture is not determined by what the best people do, or the worst. It is not what the outliers do that is not acceptable. It is entirely possible that an otherwise acceptable safe culture is being masked by what a small but not insignificant number of outliers do. Were that to be the case, there is not a culture problem, but rather a performance problem limited to subset of followers.

CULTURE: GROUP BEHAVIOR

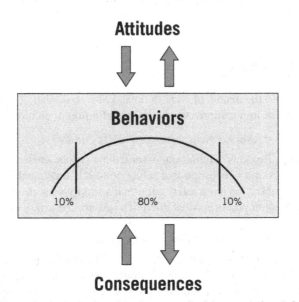

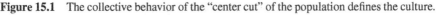

Figure 15.1   The collective behavior of the "center cut" of the population defines the culture.

3. In practice there's not usually one culture, but a collection of sub-cultures.

An assessment of the culture is typically based on the belief that everyone in the organization behaves the same way. Real life is never like that. That said, for a crew or department or even site, behavior may be tightly clustered and variability small. But consider a large industrial firm with multiple divisions and a string of operating sites spanning the globe. Can it accurately be stated that the safety culture is consistent across the entire company?

In the most poorly performing cultures, there is almost always an island of excellence to be found in a sea of mediocracy. At the site level, it's commonplace to find significant variability across functions—like between operations and maintenance—departments, shifts and even between days and nights. It's easy and tempting to lump everyone together—and sometimes impossible not to—but the better the safety culture is understood for what it is—the add up of sub-cultures—the more likely there is to be a successful change effort.

4. Understanding the variability in the safety culture changes the frame of reference of every leader.

The primary audience for *Alive and Well* is the front-line supervisor, who on the matter of culture must feel the challenge is far above their pay grade. As the front-line leader looks at the situation, it may not even be apparent that there is a culture: they see the individual behavior of each and every crew member who the leader knows by name.

It is at the level of the manager and executive that the culture problem becomes more apparent: they see "what most people do, most of the time." —the common practice across a large group.

Still, by understanding that sub-cultures likely exist, it is entirely possible that the culture challenge is, first and foremost, a problem with a particular sub-group. That means the culture change can be within the purview of an individual leader.

5. Culture is observable and measurable.

Defining culture to include assumptions, beliefs, values, norms, and climate creates a significant failure trap: in one way or another, a leader is forced to attempt managing and changing attitudes, as culture is found in the five and a half-inch space between the ears of followers. A leader who is of the view that a change in belief must precede the change in behavior has to accept "attitude change" as a principal strategic element in the process.

As explained in Chapter 8, changing the attitude of a follower is well-near an impossible task. The problem starts with being unable to know what the attitude is that needs to be changed. If that's not hard enough, then comes the task of actually changing the minds of followers, getting them to think the way the leader wants them to. That's why the common approach to culture change is a failure trap.

Behavior has none of those difficulties. If you want to know what the behavior is, all you have to do is observe. Just make sure you're observing the normal behavior of most people. Don't make the mistake of measuring the culture by what those at the extremes are doing. Good behavior can be found in the worst of cultures; the best of followers aren't always perfect.

In other words, you need data. A lot of data. Specifically you need data on the cultural dimensions that describe your safety culture.

6. Only certain dimensions need to be changed.

As a practical matter, some behavioral dimensions are likely to have a far greater impact on safety than others. That's the logic of Pareto's Law: the 80/20 rule. Understanding that to be so—and identifying the few that matter—significantly reduces the scope of the change effort. To change the culture in a significant way, it's not necessary to change every follower's every behavioral dimension. It will fall to the leader driving the culture change to determine what dimensions matter to the change effort.

7. Changing culture means changing group behavior.

It's now all so obvious: culture is what most people do most of the time; there-fore, changing the culture requires changing what most people do most of the time. That means moving the masses.

Start by doing the math: picture one CEO taking on the rest of the organization. Yes, it is the CEO, but what about the numbers? Or the distance—geographic, organizational, and social—separating the top executive from the masses? How does the CEO effectively influence followers across and down the organization hierarchy in a global operation? How does the CEO know the progress that's being made (or not) in changing the behavioral dimensions across the organization that directly relate to followers going home safe?

As powerful as the CEO is, successfully changing the culture requires a lot of good help – from managers and front-line leaders.

YOUR SAFETY CULTURE

You're living with a safety culture. James Reason had that part wrong, but he was right that a culture *of* safety is as rare as the state of grace. How close is your safety culture to being a culture of safety?

Here's where being a front-line leader or manager gives you a huge advantage over the CEO. If you want to know what that culture is, all you have to do is pause for a moment and look around. What you see is the culture. When it comes to safety, the critical behavior dimensions are easy to observe. You just need to decide exactly what they are. Then you look: for example, at how many are stopping their truck at the gate, putting the vehicle in park and getting out to safely badge in; what happens at the point where a Work Permit is executed between operations and maintenance. Or, how carefully is the paperwork being filled out; how effective are the safety meetings; how good is the safety training.

Be careful not to jump to any conclusions based on a single data point; there will always be variability. Culture is the summation of hundreds of small acts and interactions taking place every day among all those working in the organization.

No matter what your company or division or site safety culture might be, as a leader you owe it to yourself, your leaders, and your followers to understand exactly what *your* safety culture is for your followers. Your safety culture might be a good thing, with the majority of people holding onto the handrails as they walk down the stairs. Your safety culture might be a bad thing, with people ticking the boxes instead of taking the time to answer questions on the job safety analysis thoroughly. In either case, good or bad, reality is what the culture is, not what you wish it were.

Wishing your safety culture were something else—or worse, believing it's something it isn't—can be perilous.

Seeing the culture for what it is seems a simple thing. It can be. But doing that is easier said than done. If you've spent your working life in a single culture—common for many front-line leaders—what you see day after day, month after month, year after year looks like normal to you. Your normal *is* the culture. If you don't have any basis for comparison, it's easy to assume things are the same everywhere else. That may not be so. On that point, getting an independent assessment of your culture—cold eyes—can prove valuable.

To repeat: you have a safety culture. As to what it is, if you are fortunate enough to have a culture of safety—it is not an unattainable goal as there are leaders who do—your challenge is to maintain it. You may have a respectable safety culture, with a few poor performers causing your performance problems. That's a performance problem to be managed—not a culture to be changed. You may not have the safety culture you want.

If so, your goal is to create the safety culture you want.

CHANGE VERSUS TRANSFORMATION

Culture is always in a continuous state of change, and it will constantly change—evolve—on its own. If you want an example, look at the pictures of "now and then." For example, compare pictures of you in school and with the way kids dress and act now. The change is staggering. Compare pictures of the crowds at a sporting event in the 1950s with today. The stadium might not have changed, but the dress and demeanor of today's fans bear no resemblance to those of yesteryear.

Those dramatic changes in group behavior didn't come about by thoughtful design or planning. Nobody in charge decided that face-painting and chanting was such a good idea that every single fan in the stands should do it, effective the next game. More likely the opposite happened: as a reaction to culture change, those in charge had to take action to halt the evolution of the culture. For example, school administrators write new rules on what students are no longer allowed to wear to class.

It's true for safety as well: how the work is done—and how the people doing it behave—has dramatically changed over time. Culture does change, and in the case of the safety culture in most organizations, the change has for the most part been for the good. That creates an option for you: be patient, and let the culture change on its own, hoping for the best.

But if you have a culture problem on your hands, as a leader—executive, manager or front-line supervisor—you don't have the luxury of waiting around for the years it might take for your culture to change on its own to become what you want—a culture of safety. Moreover, you shouldn't be willing to trust the direction culture will take on its own with no direction from you. You're left with no choice but change the culture to exactly what you want it to be—and you need to change it on your watch.

That's *transformation*: a rapid change intended by design. Transformation implies both speed and direction.

It is *cultural transformation* that has proven so brutally difficult for leaders. The culture of the fans in the stands in stadium attending a sporting event slowly evolved over a period of years to become what it is today. That's what differentiates evolution from transformation. With evolution, you get a random walk to a place you might not like the look of when you get there. Transformation produces a predetermined result achieved at a rapid pace.

Transforming the safety culture begins with understanding the nature of culture. The first step in the process is to establish an understandable definition of culture, part of creating a common vocabulary: culture is what most people do, most of the time. This understanding creates a succession of useful insights: for example, culture can be observed and measured, you have a culture, for better or worse and change means moving the masses, but not everyone has to be moved on everything.

Understanding that explains the critical need for the clarity of vision as to the desired change: a clear and detailed picture of the culture a leader wants to have, defined in terms the rest of the organization will understand and can take actions to help actualize. Then comes the really hard part: causing the culture to change in the desired direction, and doing so with speed. That is the transformation.

Contemplate doing that, and you can begin to appreciate how rare it is for group behavior in an organization to rapidly change by virtue of a strategy driven from company headquarters.

Try as they will, leaders find that the culture proves brutally difficult for them to change, even when it's the senior leader driving the effort. On the other hand, if you're working as a leader in the middle of the organization, the reverse seems true— the culture can't be changed, let alone transformed, without top management leading the effort.

So, in the belief that change starts at the top of the organization, an effort to change the culture should start by changing the group behavior of the top executives. All too often top management won't change, doesn't see the need to change, or isn't up to the task of making change happen. Change never happens.

Meanwhile, on the front line of the organization, where most of the significant behavior that is described as the culture is found, leaders deal with the culture every day. Rarely do these leaders believe they have the power to change the culture. They wait in the hope that *their* leaders will do that for them.

As a result, nobody feels any power to change the culture. It's the great paradox of culture change. Leaders at every level are frustrated, but no leader feels capable of effecting change.

But if the culture—particularly the safety culture in an industrial facility—is to be transformed—changed rapidly in a desired direction— middle management actually holds the power to make that happen.

A ROADMAP FOR TRANSFORMATION

If you're a leader and you know exactly the safety culture you want for your organization, how do you make the transformation happen?

Sadly, there is not a set of simple, repeatable steps which, if taken, guarantee successful change. There simply aren't enough successful transformations from which to define a set of best practices. Moreover, no two organizations are the same; no two organizations start from exactly the same place; the desired behavioral dimensions of the safety culture vary.

That doesn't suggest you don't need a roadmap for your transformation: you absolutely do. But you'll have to come with your own unique roadmap: the simple, practical steps you will take, and the measures to know the progress you're making. A roadmap that fits you and your operation; where you are, and where you want to go.

That said, there are principles and techniques for changing group behavior that can and should be incorporated into your roadmap, and successful cases with useful lessons to serve as examples.

1. The first principle is clarity.

You must have a clear picture in mind of the behavioral dimensions you want to see present in the culture you want.

Leaders in all walks of life are onto something when they're frustrated by the culture. The cultural sore spot might involve safety—not following the safety rules, not reporting near-miss incidents, mindlessly ticking the boxes on the job safety analysis—or social issues—littering, low voter turnout, or street crime—but all are legitimate manifestations of a culture in need of change.

But those who want to transform the culture then make a tactical error: attaching a label to the problem and directing the culture change at the label. A label is a shorthand means of expressing an idea. In the case of culture, labels can misdirect and even subvert the transformation effort. Here's an example.

In the aftermath of NASA's *Challenger* accident, the poor communications up and down the management chain got its fair share of the blame. Engineers and scientists had knowledge about technical performance problems with the equipment—leaking O-rings, for example. But those at the top of the organization weren't aware of those problems.

NASA described the flaw as a communication problem. That label set up one key element in NASA's effort to change the culture: improving how well people listened to each other. To an outsider looking in, the real cultural problem was better described as the following: "The people at the top of the organization don't want to hear any bad news that might stand in the way of the mission. If you're the messenger bringing bad news, be prepared to be shot!" One aspect of the safety culture in that organization was to avoid bringing up bad news and to express a "can-do" attitude. NASA is hardly alone in having that slogan as one aspect of its safety culture.

The ensuing training in empathetic listening skills changed that part of the culture not one iota. When *Columbia* was launched in 2003, some scientists and engineers who suspected there might be a problem with the orbiter's wing got anything but a sympathetic hearing when they brought their concerns to the attention of their managers.

Labels fail in a more fundamental way. There's a tendency to sum up the desired culture in a simple phrase like "culture of compliance" or "can-do culture" or "safety is a core value." That's nice for a start, but it's the details that count in transforming culture. It takes far more than a catchy phrase or buzzword to create culture. You have to be able to spell it out in great detail to realize the culture you want. Otherwise it's left up to the people in the organization—the existing culture—to determine what the different behavior looks like. They'll be off every time.

Find an organization with a strong and positive culture, and you'll find management that understands exactly what it wanted, and has made achieving it a critical goal. That's true for landing airplanes on aircraft carriers, friendly service at McDonald's, and, yes, NASA's superb effort at rescuing a tumbling spacecraft.

Lost in the rush to invest intellectual capital in the study of culture is the simple fact that the leaders with the most success in creating the culture they wanted (i) didn't use consultants, (ii) didn't commission studies, and (iii) didn't have a mission, vision, and values off-site meeting to decide what to do. Walt Disney, Dave Packard and Bill Hewlett, and Irenee DuPont created powerful cultures. They were clear about what they wanted to see happen in the organizations they ran. They successfully converted that view into reality with the right mix of actions that produced the results they wanted to see.

They started with clarity about what they wanted.

Walt Disney and his Disneyland and Disneyworld serve as wonderful examples. What is it that makes Disneyland so special? Anyone who has been there can describe the Disney culture in detail: friendly service, manicured grounds, immaculate housekeeping, and quality artwork and sound in the exhibits.

You can be sure this is exactly what Walt Disney had in mind. Surely it's exactly the picture he painstakingly painted for those in charge of building and operating Disneyland, a culture that has flourished long after his passing. Disney did start out life drawing cartoons, which combine imagination and vision in a very vivid and powerful – way. What a great tool for a leader to become clear about the culture he or she desires.

If you are a senior leader, you need to be able to do the same thing for the safety culture you're trying to create. Describe how the culture looks to someone in the thick of it—a customer, a new employee, or a key associate who performs critical work for the organization. If you find that difficult—for some leaders it really is—try asking yourself what kind of behavior you would see on following important dimensions if you spent a day out on the shop floor:

- How well the safety rules and procedures are followed
- What happens when someone is seen working unsafely
- How often near misses are reported
- The condition of tools and equipment
- How often people make safety suggestions, and what happens when they do
- What goes on during the safety meetings
- The amount of time leaders spend with their team members
- What leaders talk about during conversation with subordinates

Write down what you want, and then explain it to someone who doesn't know what you're talking about. If you can hear it back the way you want it to be, you are on the right track. If you can't describe what you want—in a meaningful and clear way—so that those responsible can help you change, build and manage your culture, "Do not pass Go!"

2. The second principle for your roadmap: instead of looking up for help, look for leverage.

Doing that isn't easy, in part because the real leverage points aren't obvious, and are sometimes even counterintuitive. When it comes to cultural change and transformation, the great tendency by just about everyone in an organization is to look to the top of the organization, thinking that's where the power to change exists. That this approach is more likely to result in disappointment rather than real change hasn't stopped people from continuing to take it. It's what the board of directors does when it appoints a new CEO, or what the voters do when they elect a new president. If you stop and think about the kinds of sweeping changes that have a profound effect on culture, they don't normally start at the top of an organization and work down. Revolutions start at the bottom or from the outside and work up or in. Trendsetters who dictate the clothing fashions that sweep the nation don't rely on the power of an organization to provide their influence over the rest of us.

In his book *The Tipping Point* Malcolm Gladwell recognized that the process of big change is counterintuitive. The conventional wisdom of change goes along the lines of "start at the top of the organization and work down" or "solve the root causes of the big problems, and the symptoms will correct themselves" runs counter to real life experience. Relatively small changes can have a disproportionately large impact. So, for example, Gladwell described what he called the "Law of the Few": a relatively small number of people are the shakers and movers who are found at the epicenter of sweeping changes.

In the world of business and industry, those few with the great impact don't necessarily occupy the corner office. The reality is just the opposite, something we've known since we were kids on the playground. In every group there are natural leaders who have a great influence over the rest of us.

When we grow up and go to work, most of those natural leaders don't find their way to the top of the organization. But that doesn't mean they've lost their ability to lead; it just means they don't have one of those high-profile leadership positions. If you're a leader and you know where you want to go—to change the culture in some very specific way—these leaders can help you get there. You know who they are: they normally stand out in a crowd. They'll do the heavy lifting of creating buy-in and support for change, and for them it might not seem much like hard work. Natural leaders can be that good.

In industrial organizations those natural leaders are often found at the front line. They might not have had the interest in education that got them into an engineering school; instead, they went to work as apprentices or operator trainees. Their leadership skills were recognized, and eventually they became front-line leaders. Most of those front-line leaders have huge credibility with the people they supervise.

3. The third principle: sell your change.

In transforming the culture front-line leaders represent a huge source of leverage. If you're going to enlist the help of these natural leaders, there are two important steps to take.

Buy-in increases dramatically when people understand the reason for the change. Take the time to sell people on the change you want to make. Changing the safety culture for the better ultimately takes you right back to the case for safety.

Remember, changing the culture means changing behavior, and therefore changing the way work gets done. If the old way of working meant taking too much risk, what specifically is the new way to do the work that requires less risk? What is the impact from that change in the way work is done?

If there isn't real change in behavior and how work gets done, the culture can't have changed.

4. The fourth principle: small things can play big.

When it comes to culture change, logic suggests that it's necessary to first find and then address the fundamental causes. There are problems with that logic. There is no telling what the root causes of a culture are. If you did know them, likely they would be monumental, far beyond the problems you can fix.

The crime problem in New York City was just such a problem. Think about the potential causes for crime in a big city: the economy, poverty, poor education, broken families, drug abuse. Those monumental problems have been attacked for years, with little to show for the effort. If you were running the city's subway system and tried to address root causes, you would just give up. Why waste your time trying to change that culture?

According to Gladwell, that's not the approach taken by the director of the subway system, David Gunn, and the man he hired as his chief of transit police, William Bratton. No great surprise that Bratton was described as one of those charismatic, natural leaders. The two decided to work on the other end of the problem spectrum: the small criminal behavior that they actually had some chance of significantly affecting. No doubt this struck some as working on solving the wrong problem, and an inconsequential one at that. Bratton and Gunn employed a different logic, known as the *Broken Window Theory*. That theory of policing holds that criminals—either consciously or unconsciously—size up the conditions of the environment in deciding where, and perhaps whether, to commit street crimes like mugging and robbing. Remove the conditions that lead to crime—the "broken windows"—and crime rates go down.

Applying the Broken Window Theory to the New York subway system in the 1990s looked like this: going after the people jumping the turnstiles instead of paying the fare, and the kids who spray-painted graffiti on the sides of the cars. Hardly the kind of problems that seem likely to transform the culture of crime on the subway system.

But they did. At the end of the decade, subway crime had declined by 75%. The culture had indeed changed for the better; the Broken Window Theory is a reasonable explanation for the change.

All of which suggests when it comes to changing culture, not only can a small number of people have a profound effect, but so can relatively small changes. Proportionality—big effort for big change—doesn't seem to apply. The trick is to enlist the right people and find the right changes that can produce the great results.

About the same time that Gunn and Bratton were working on the culture of the New York City subway, another born leader, Charlie Hale, was in the midst of changing the safety culture at a chemical plant 30 miles west of Manhattan. It was an old plant, dating back to the 1930s, the place where the plastics business was invented. But that was a long time ago, and the safety culture there seemed stuck in the thirties. And Charlie was appointed as the site manager.

Anybody who ever worked for Charlie would tell you he was passionate about safety. His attempt at changing the culture in this plant started with the stop signs. He'd flag down someone who had run a stop sign inside the plant, and ask, "What part of the word *stop* do you not understand?" A great, natural leader like Charlie could pull that off. It helped that he would heap praise on those he found doing the right thing, and it always came across as sincere.

Though Charlie didn't call it the Broken Window Theory, he was following exactly the same logic. Start with something small but significant: a winnable battle. Then build that into something else. It wasn't long before the safety culture at this site was transformed.

When Charlie left the job he was given a going-away present: a bright, shiny stop sign personally autographed by everyone at the plant. He hung it over his desk at world headquarters.

5. Seize the moment!

One of the fundamental frustrations that leaders have about culture is how quickly it is cemented in. Hire a new employee and in a matter of weeks he behaves just like the people who have been working there for years. Open up a new department and it isn't long before the housekeeping in the place is just a bad as in the rest of the plant. In the case of the subway system, you can be sure the newest cars were the favorite targets of the graffiti artists!

That's true. But what is also true is that there are times and places in which followers are most receptive to the influence of their leaders. Those are the Moments of High Influence. In those Moments the culture may not be ready to surrender, but they present leaders with an opening, a place to begin to have some impact.

New employees joining the company, the department, the crew. A change in leadership; a leader's promotion or transfer. A crisis, in which people may see what is really important to the leaders—taking care of the business or putting safety first. A serious injury, after which people are forced to think about the implications of safety on human lives.

The challenge for a leader in those moments is simply to act as a leader in a way that advances his or her cause. If creating a culture of compliance is the goal, it makes a very powerful statement to insist in a production crisis that everyone follow the safety rules, even if it slows down the response.

On that principle, there is no finer example to be found than that of Paul O'Neill, who took full advantage of his Moment of High Influence when he was introduced to the financial community as the newly appointed CEO of Alcoa. More than a decade later he gave a first-person account in a speech at a safety conference that became the basis for a chapter in Charles Duhigg's bestseller *The Power of Habit*. Duhigg called the chapter "The Ballad of Paul O'Neill." For anyone interested in changing safety culture, this is a must read.

Mr. O'Neill began the session by pointing out the emergency exits and how to safely evacuate the room in the event of an emergency, and he proceeded to spend the entire press conference talking about workplace safety. He announced his intent to make Alcoa the safest company in the world. The investment analysts thought O'Neill had lost his mind and the Board of Directors had made a huge mistake. Thirteen years later, O'Neill proved that initial judgment completely wrong. Safety performance had improved dramatically, as had business performance and share-holder returns. It wasn't until O'Neill had long since retired that he revealed his strategy: to use safety behavior as what Duhigg called a "keystone habit" from which to change not just what people did to work safely—behavioral dimensions—but ulti-mately how they went about doing their jobs, changing their collective behavior for the better.

In a word, that's culture.

Paul O'Neill didn't just recognize his Moment of High Influence, he showed up with his roadmap for transforming the culture. The stuff of genius!

CONCLUSION

The fact that culture, by its very definition, is so embedded in the fiber of an organi-zation is what makes it so difficult to change. But that same inherent quality means that a strong and positive culture—one directed at important results—can be an overwhelmingly powerful force. It dampens the ups and downs in the life of an organization. Culture operates like a gigantic flywheel, stabilizing and normalizing behavior. That can work to better—not worsen—performance.

The methods proven successful in changing culture are in some important ways counterintuitive. The normal rules of logic—cause, effect, and proportionality—don't hold up well. Follow the conventional wisdom and you're more likely to fail than to succeed. Better to heed Gladwell's advice: "with the slightest push—in just the right place—the world can be tipped." Instead of looking up, better to look down to find the leverage, think "small but significant" and seize the Moments of High Influence.

Finally, no discussion of cultural transformation should overlook the most basic and simple impetus of change...how the leader acts. No leader can expect the rest of the organization to change its behavior—and, therefore, culture—to something different from what the leader practices.

It's called Leading by Example.

CHAPTER 16

INVESTING IN TRAINING

> Training is probably the least effectively used management tool in industry and business.
>
> —*William McGehee and Paul Thayer*

Every year industrial companies invest billions of dollars training their people; and safety training represents a considerable chunk of that expense. When put into practice, all too often the execution of that substantial and important investment in safety training looks like this.

1:10 p.m., Department Training Room

An hour into the annual refresher training on the company's life-critical safety procedures, the front-line supervisor surveys the scene from the back of the classroom. What the leader sees is not a pretty picture.

The lights have been dimmed and the latest in what seems like a never-ending show of PowerPoints shows on the screen. The course instructor, facing the screen, drones on.

As to the course participants—all followers of this leader—the effect is totally predictable—and looks even worse.

The signs of boredom are unmistakable: there's a side conversation about the upcoming hunting season. One student is engrossed in a crossword puzzle, another is scrolling through text messages. At least those four are still awake. As for the rest—well, daydreaming might be the kindest characterization as to their level of engagement.

Who could blame them, the supervisor thinks. It's not like they haven't heard this stuff so many times before.

It's not hard to feel sympathy for the safety department staff member, handed the unpleasant duty of teaching refresher training. They always give this assignment to their newest staff member, whose degree in safety management hasn't prepared them for this kind of duty. It's a good bet they were handed the slide presentation and pointed in the direction of the training room.

The supervisor breathes a sigh of relief, thinking "Well, at least this isn't my problem." But what a waste of time, just so somebody in senior management can be shown a report showing, "Staff 100% trained and qualified."

But trained and qualified for what?

Alive and Well at the End of the Day: The Supervisor's Guide to Managing Safety in Operations, Second Edition. Paul D. Balmert.
© 2023 John Wiley & Sons, Inc. Published 2023 by John Wiley & Sons, Inc.

This familiar scene leads to the obvious conclusion that training can't be all that important to safety, to the business, and to the leaders running the business. Is that so? Should that be so?

Of course not.

Still, a chapter about training might strike you as being out of place in a book about leading and managing safety performance. It is not. As to the explanation, we'll begin the examination of the subject of training with this question: What is the most basic and fundamental principle of working safely? Asked another way: What is the one thing that matters most to working safely?

It's a question for which the correct answer should be on the tip of the tongue of every leader. It should be one of those "Every leader knows the most important thing is…" kind of moment. Rarely is that the case.

Think most important thing is caring?

If you think caring is the most fundamental principle in safety, you need to read Paul O'Neill's account of the turnaround of safety performance at Alcoa. It can be found in a speech he gave at a safety conference in 2001; later it became a chapter in Charles Duhigg's *The Power of Habit*. O'Neill's message is so powerful, it should be required reading for every industrial leader.

When he showed up at headquarters as the newly appointed CEO, he quickly found out everyone cared to the point of shedding tears over accidents, but did nothing to actually change things for the better. "Caring is not enough. Caring is not nearly enough" O'Neill said to his followers, and then set about to change safety habits, which he firmly believed was the effective means to seeing to it that everyone did go home, alive and well at the end of the day.

You would be hard-pressed to find a leader anywhere on the planet who doesn't care about the safety of their followers. Still, many struggle to manage safety performance well. O'Neill was absolutely correct: caring is not nearly enough. If it were, every follower would work safely.

Is it the elimination of hazards?

A worthy goal, but matter how hard you might try to get rid of them, there will always be hazards. As you now know from reading the chapter on managing risk, hazards are the things that can harm people. Anything can be a hazard; the only way to eliminate all hazards is to remove all the people working around anything that can harm them. That goes for you, too.

That's not going to happen.

Is it having procedures to manage hazards? As important as they are, there will never be a procedure covering every hazard. Moreover, for all the procedures you already have, if they're not being followed faithfully and completely, they're little more than pieces of paper.

So what actually does matter most to safety?

The answer is stunning simple: knowledge. Knowing what the hazards are, and the means to keep from being harmed by the hazards.

In practice, knowledge represents the first line of defense in safety. In a real sense, it is the single best way to protect followers from harm. What they don't know can hurt them; if they know they can do the things needed to keep themselves from harm.

Training functions to cause the successful transfer of the knowledge necessary to enable people to work safely. It is that simple, and training really is that important. But the effective transfer of knowledge by training is neither simple nor easy.

KNOWLEDGE

Consider everything followers need to know to be able to perform their jobs safely. First, there's the technical know-how related to operating, producing, maintaining, and moving the product. There is the information involving performing the work they are assigned in a way that meets safety requirements. These requirements begin with the basics of personal protective equipment and progresses through following life-critical procedures such as entering a confined space. In between there is knowing about everything else: inspecting, evaluating, reporting, investigating, and documenting.

For you to understand just how much knowledge there is for your followers to know, you might try making a list of everything any one person on your crew or in your department needs to know to work safely. That is something every supervisor ought to know, not just for one follower, but for every follower, and for every job. Few do. In theory, you could look it all up—it should be written somewhere—but you'd probably have to look in so many different places that it's just about impossible to find. Besides, you're a very busy leader; who's got time for that?

In principle, nobody should ever get hurt because of a lack of knowledge. The necessary knowledge to work safely is under the control of supervisors and managers. Leaders know the hazards and the procedures to work safely, and are responsible for the training and qualification of those assigned to do the work. Work should be assigned only to those with the knowledge to work safely. No one should ever perform work they do not know how to do safely. This is true no matter who they are or what they might do.

In practice, what people don't know has hurt them. There are many instances, big and small, where the lack of knowledge was a significant factor found in the root cause of the event. One of the biggest was the loss of a nuclear reactor at the Three Mile Island generating station. At the root was a lack of understanding of the reactor process by the control room operators. The loss of the reactor resulting from this event ultimately cost the owners of the facility twelve billion dollars. The training and certification process now required by the Nuclear Regulatory Commission for reactor operators confirms that knowledge was lacking.

At least nobody was hurt at Three Mile Island.

The same can't be said for the accident at a large refinery in Texas City, Texas in 2005. There 15 people died in an explosion in an operating unit. In the opinion of one of the agencies that conducted an independent investigation, the United States Chemical Safety Board, "Inadequate training for operations personnel, particularly for the board operator position, contributed to causing the incident. The hazards of unit startup…were not adequately covered in operator training."

That's an illustration of the critical need for the successful transfer of operating knowledge. Here's an example of a failure to transfer knowledge about procedures.

In 2017, at an electrical generation station, four people were killed and two others seriously injured when the contents of a tank they were working on unexpectedly released. One of the two injured later died. Found to be at the root of the problem was the procedure to perform what was clearly a very hazardous task. There was an energy control procedure and it had recently been reviewed and updated. But those working on the job knew hardly anything about that. In the immediate aftermath of the event, eight of the nine people interviewed said they never saw the procedure. Finding it required searching the company's intranet; there were no copies of the procedure to be found on site. Presumably that explained why the procedure wasn't followed and the event caused.

You would like to think the lack of knowledge as to the proper procedure to follow to safely perform the work would have stopped the work from being assigned, in this case, to a contractor. You would also like to think the absence of knowledge as to the procedure would have stopped the contractor from undertaking the work. But it did not.

Killed were employees working for both owner and contractor.

COMPLIANCE AND CHANGE

Since there will never be a procedure or rule for every potential hazard, as a leader you depend on your followers to have sufficient knowledge about what can hurt them to be able to recognize hazards and take appropriate precautions to protect themselves from harm. Your operation employs a huge array of safety policies and procedures to safely manage many hazards. For these, you are dependent on knowledge for compliance. New and revised policies, procedures, programs, and standards arising from continuous improvement, the response to the need to raise the bar on performance, to comply with external requirements, and the application of lessons learned from serious incidents drive the need for the successful transfer of knowledge in order to make change happen and compliance to occur.

Ask leaders the biggest challenges they face in getting followers to work safely, compliance and making change happen are guaranteed to show up on their list. Successfully managing both starts with effective training (see Figure 16.1).

Making change happen is not a simple matter of "communication". A change in how work is done often necessitates the need for training on how to do the work properly. At the point of execution, you never want to hear "I didn't know" as the explanation for the failure to comply (and, true or not, you'd rather not hear as the excuse, "I forgot.")

The process of gaining full compliance with the rules begins long before you get to the point of enforcement. The first step in the march to full compliance is to establish that everyone expected to comply with the rules knows and understands the rules. How can you expect people to comply with requirements they neither know nor understand?

People must be successfully trained. It all sounds so simple, but if you picture what was going on in the training room when the safety rules were being taught, you can appreciate the difference between theory and practice.

COMPLIANCE REQUIRES EFFECTIVE TRAINING

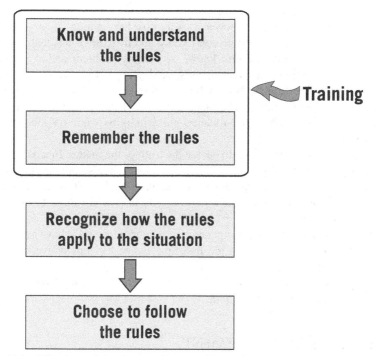

Figure 16.1   Effective training is essential to meet the first two conditions necessary for full compliance.

As a leader—front-line supervisor, manager or executive—training deserves your serious attention.

TRAINING ISN'T ALWAYS THE PROBLEM

Confusing matters are situations like this.

Joe knows how to wear his PPE properly. He is a model of good behavior and compliance when he's working in the shop. That's because he knows the supervisor will see him there—and make him put on his equipment if he's found in a state of noncompliance. But when he's out on a field assignment it's an entirely different matter. The supervisor's not around to enforce the rules, and he knows that.

So Joe takes advantage of the situation, and doesn't comply.

One day, working on a field job, he crosses path with his supervisor, who finds him in noncompliance.

"Joe, I see you're not wearing your hard hat and safety glasses. What's the story?"

Joe doesn't miss a beat: "I didn't know I had to wear them out here."

Leaders deal with problems like this on regular basis. When a follower not following the rules explains, "I didn't know it was required" or "I forgot" it sounds like a training problem. It might actually be a training problem. If so, the proper solution would be to send the follower back to refresher training.

There are times when people don't know, and there are times that people who know perfectly well say, "I didn't know," A training problem and a performance problem are two entirely different matters, calling for entirely different solutions. Yes, knowledge is essential, but sending a follower who knows the rules to be retrained is a waste of their time and your money. If the person knows better, the proper way to deal with a performance problem is to work on motivation and consequences.

It falls to you as the leader to decide the truth of the matter. In a perfect world, your followers would tell you the truth, but you know that's not how things always work in the real world. The big advantage you have, particularly if you're the front-line leader, is that you know your followers; you have a history with them. You'll likely know the truth.

If Joe has been a longstanding member of your crew, you'll have a track record to make the call. "Joe, this isn't the first time we've had this conversation about following the safety rules on field jobs."

The training problem is too serious to compound it by mislabeling a performance problem as a training problem.

TRAINING IS AN INVESTMENT

Rather than a cost of doing business, training is better viewed as an investment in human capital. Seeing training in that light is one important step in changing performance. Training should be undertaken with the expectation there will be a return on that investment; new knowledge will produce benefits.

Like any business investment in capital goods and equipment, the cost of training can be calculated. The payroll and opportunity cost for time spent in the classroom is likely the biggest single expense; there's also the cost of teachers, training rooms, and class supplies.

While everyone knows training represents a considerable investment of the scarce time and human resources of an organization, the significance of that investment is often lost in the process of running the business. Unlike sales or depreciation, the direct cost of training courses isn't added up and shown on the income statement. The indirect cost of delivery and administration and the opportunity cost have little to no visibility.

Unless they ask for a calculation, or someone does it for them, those in senior leadership positions seldom appreciate the size of the investment in training being made in *their* business. Were they to know, they well may take a more active role in managing and overseeing the investment—and demand to see evidence of the return.

That would be very beneficial to improving the training process.

More than two decades ago, a calculation was made for the attrition expected for process operators employed by a global chemical manufacturer. In the chemical industry, process operators play a vital role in safe and efficient operations and their training is a multi-year effort. Even using conservative assumptions, in a mature business, operator turnover was still significant. Combining forecasted attrition replacement with an estimate of the time and expense to be spent to bring new hire operators up to the competent level produced an estimate of the total investment in new operating staff.

Would it surprise you to learn the investment in training new operating staff across this company would total more than $100 million over five years?

By way of comparison, that expense was roughly the equivalent of a major capital project to build a production unit. A capital expansion of that magnitude would require a capital proposal and thorough review of the proposed investment at the highest level in the company. Before deciding to approve a project, executives would demand to see the business plan, examine the costs carefully, and evaluate the estimates for the return on the investment. Once the project is approved, a project manager would be appointed to oversee the design and construction. Just as importantly, those managing the project would be held accountable for seeing to it that the project is completed on time and on budget, and the investment performs as advertised.

But for a comparable investment in human capital in the form of training, no executive would even think about that level of project management and oversite, let alone insist it be put in place.

The authors of *Training in Business and Industry*, William McGehee and Paul Thayer, understood the practice. "Upper management is rarely aware of the dollars-and-cents implications of training outcomes. Consequently, training is probably the least effectively used management tool in industry and business."

They wrote that in 1961.

McGehee and Thayer summed up the potential return on the training investment: "The effectiveness of achieving...organization goals will depend, in a significant way, on the nature and efficiency of the training employees receive for their assignments." In the twenty-first century, what's the "nature and efficiency" of training? Lecture by PowerPoint or computer-based training? What kind of return can reasonably be expected from that approach?

Of course, McGehee and Thayer also noted that the benefits from training—and the costs of poorly trained employees—were just as likely to be underappreciated. Training is seen as "something you have to do." The benefits (or the lack thereof) are not easy to quantify. For example, how do you put a number to the benefit of having no injuries or incidents caused by a lack of training? To be able to do that would require someone to keep count of all the cases where lack of knowledge was a significant causal factor, and then track the improvement.

On the other hand, perhaps someone should. There are ways to get an estimate as to the contribution made by training. The trick is to know where to look.

Training expert Don Kirkpatrick identified three very useful places to look for information:

- What has the student learned?
- How has the learning affected the student's job performance?
- How has the student's job performance impacted organization performance?

As to how to go about applying Kirkpatrick's model to safety and training, it could be as simple as doing what amounts to a "root cause of root causes" by looking at safety investigation reports collectively rather than individually. Where that has been done for a division or a company, the findings have spoken powerfully as to where more training is needed—and is not.

Kirkpatrick suggested a fourth place to look: at training course evaluations. For the teacher who teaches a course, student evaluations are a useful way to do the job of teaching better. But, for those footing the bill for the training, improving class-room teaching is the least of their problems. Wiser to go after the other three evaluation questions first.

"Price is what you pay. Value is what you get." —sage advice from Warren Buffett describing his investment philosophy. He could have been speaking about investing in training.

WHO'S PROBLEM?

Ownership—or perceived lack of ownership—is a significant factor in "the training problem." You would think nobody would be more dissatisfied with the return on the investment in training than senior management: it is their money being spent. Yet seldom are they the ones complaining about training; usually it's the students sitting in class, muttering under their breath "This is an hour of my life I can't get back."

If you're a senior leader or executive, reading that should be all it takes for the motivation and impetus to look into the training problem. If nothing else, your good followers deserve better. If you're of the opinion that training as characterized in the illustration is not the norm in your organization, there's an easy way to prove your view correct. Sit in on a mandatory safety training session and see for yourself. Or ask the participants' leaders at the front line what their view is as to the effectiveness of safety training. If they tell you there is problem, then consider the message poor safety training sends as to management's commitment to safety.

On the other hand, if you're a front-line supervisor, the training problem may well be one problem that really is not *yours*. You've got more than enough other problems to deal with, and you probably think you couldn't do anything to fix the training problem even if you wanted to. Easier just to send your people to training on new safety policies and procedures, corporate safety initiatives and information systems, refresher training on the stuff they've been doing for years, and training on your customers' procedures and requirements, and then tick the box that says "trained". If the training was a waste of time, at least it wasn't your decision.

That's one way to look at it—but not necessarily the best way.

Yes, senior management may be dictating the training take place, and yes, the training department may be in charge of course design or delivery. Neither may be asking you what you think should be done about training your people. You could point the blame at the Training Department, but for openers, they aren't necessarily the ones responsible for producing and delivering safety training. Even when they are involved, they depend on Subject Matter Experts with expertise in the specific course content, which is often highly technical. Even when the expert works in the safety department, their technical expertise doesn't often lie in instructional design and adult learning theory.

As a practical matter, a lot of safety training is developed by someone tasked with the assignment who has good intentions but little training expertise. Hence, "death by PowerPoint."

All that may well be true, but as the front-line leader, those people falling asleep in the class are *your* people. *You're* the one responsible for what they know and don't know. You're the one accountable when something goes wrong. After a serious injury, knowing you weren't the one doing the poor job training will be of little consolation to you. Bottom line: you don't want to see a follower of yours hurt because they were not properly trained.

The better way to look at your role in the process is this: you are a very important customer of the training process. What the customer thinks really does matter.

So, speak up.

A PERFORMANCE PROBLEM

As a supervisor, you deal with matters related to the effective transfer of knowledge every day. "Is Ann qualified to sign a work permit?" "Did Joe know how to wear his respirator properly?" "Who knows how to get this accident information into the computer system?" An advanced degree in adult education isn't necessary to successfully deal with the training problem: a little bit of common sense and clear thinking will serve you well.

The first question to ask about training is, "How important is training to the success of the business?" You now know the answer.

The second question is, "Do we know how to train people well?" Yes, "we" do.

Since we've all had good teachers, taken good courses, and benefitted from good coaches, collectively we know how to teach well. Recent scientific research into the functioning of the brain has given us great insight into how we humans learn. Ironically, that knowledge only serves to reinforce the teaching practices of great teachers like Socrates. Twenty-five hundred years ago he figured out the power of teaching by means of asking questions. It's still the best way ever invented to teach.

Learning experts can give you a complete explanation as to why his technique works so well. But you don't have to have a PhD in neuropsychology to appreciate his genius; try asking someone a Darn Good Question, you'll see the evidence firsthand.

We also know what doesn't work: putting an ineffective trainer in front of a class, reading PowerPoints, finishing with, "Any questions?" That's no way to learn. But we regularly choose—albeit unconsciously—not to put what we know into practice, and go along with what we know doesn't work.

There's an old training adage: "If you put a gun to their head and they could do it, it's not a training problem." Put a gun to our collective heads and we can train well. But we seldom do. Why? Because delivering good training isn't easy.

That makes training a performance problem.

THE THREE Ts OF TRAINING

Changing training performance requires seeing the need for change, and then improving the process. At this point, you've read more than enough to convince you training needs to be changed—for the better. If you're a top executive, you should throw down the gauntlet and challenge those responsible for training to improve. They are spending your money and you should not be satisfied with the return you are getting. If you're a front-line supervisor, you can act like any good customer, and insist those doing the training for your followers do better.

Actions like these may well be what it takes to begin changing the training culture. Remember our definition of culture: what most people do, most of the time. Common practice. Your operation has a training culture; most likely it is not what it needs to be.

At the other end of the change spectrum is you and your training performance. Do you run training courses yourself? Do you directly supervise followers who run training courses for you? It's almost certain your answer to one or both questions is yes. If so, improvement only requires you to change yourself and your followers: that you are totally capable of doing. No outside intervention is necessary.

How do you improve training, starting with *your safety training*?

Every student deserves to be taught well, and just about anyone is capable of teaching reasonably well. You don't need a PhD in adult learning theory to be good at teaching. The hallmark of good teaching and effective training courses isn't the content: the content is always the content. It's the *process* of teaching that makes the difference.

The definition of *teach* is "to cause to know a subject." Take note there two verbs: *cause* and *know*. Only one noun: *subject*. A good teacher creates not just knowledge, but also the motivation to learn. To meet both objectives, the training process must start with the student, and then work backwards to determine the design and delivery. An old-fashioned notion, one with a different result in mind than much of what gets labeled as "training."

As simple as the principle of designing around the learner might be, that is not how many who design and deliver industrial training courses approach their work. Designing training courses—particularly the one-of-a-kind classes for safety, such as "training for the new equipment isolation procedure" —follow the familiar model of "this is how we always train." In other words, a classroom lecture built around a set of PowerPoints, read by the teacher. Followed by, "Any questions?"

Sound familiar?

That design makes it easy for the teacher—the teaching notes are right up on the screen—but tough on the learners. Being handed that design handcuffs the teacher: "Use this set of PowerPoints to train your crew."

There is a better way: follow the Three Ts of good training: *timing, technique,* and *teacher*.

Timing: There really is a best time to teach any subject. That's when the student is ready to learn and will immediately use what is taught.

Technique: There is a much greater range of techniques that a teacher can employ to teach any skill or knowledge, from asking questions to running laboratory experiments. The key to effective design is to choose the best technique for the *student* to learn—not the teacher to teach.

Teacher: While it is the responsibility of the student to learn, a good teacher can spell the difference between simply passing a test and really mastering the subject. Anyone who's ever sat in the classroom as a student knows that.

When it comes to training, these three factors make the difference. They're basic, simple, really just common sense. Common sense, yes, but rarely common practice.

Simple may be simple, but putting simple into practice is anything but easy. Delivering training *when* the knowledge is needed, using methods that work best for the *learner*, and *developing* good teachers requires hard work on the part of those responsible for training. By comparison, doing it the way we always have, and ticking the box is easy.

Faced with a choice of either "do it well" or "tick the box," most leaders collectively choose to pass on the challenge. Training over, management can say, "They're trained and qualified." Trained, perhaps, but as Charles Kettering put it, "Knowing something and understanding it are not the same thing."

In practice, understanding is what matters.

THE FIRST 'T': TIMING

There's more to timing than first meets the eye. The best time to learn something is when someone is ready to learn and needs to know. In business, though, training takes place in the context of an ongoing operation: the customer must be served, the product made, the work done. There's is always a trade-off between the best time and the most convenient time.

The cost of taking people away from their productive work and putting them in a classroom—known as *opportunity cost*—is huge. When their work is covered by replacements, it may require overtime. Student cost is normally the biggest single expense of training. Add in the costs of the instructor, the room, and the logistics to support a class. Perform a "true-cost accounting" of a training course and you'll find it's a lot more expensive than it appears.

The high cost of training explains the popularity of computer-based training programs that don't require a class, classroom, or teacher. If the student can perform their regular job while taking a required training course—say, sitting in a control room, operating the process, and turning the pages on the CBT module "learning" the opportunity cost of training is reduced to zero.

But the image of an operator in a control room watching PowerPoints on a computer instead of process controls doesn't build confidence.

Computer-based training also solves the other big problem of timing: delivery when the training is needed and will be used.

It requires no genius to understand the benefits of delivery at the point of immediate use. Adults learn best when they are ready to learn, and they are never more ready than when they need the knowledge. Delivery at the point of immediate use also solves the retention problem: we humans forget very quickly that which we do not regularly use. Research on retention suggests significant deterioration of the ability to precisely recall and apply information within two or three weeks of learning it.

In theory, computer-based training can solve the timing problem. Of course, that assumes that computer-based training is an effective way for someone to learn. But much of the safety knowledge that needs to be transferred falls outside what can effectively be taught by computer: when it comes to putting out a fire, there is no substitute for practice beforehand holding a real fire extinguisher.

The best time to learn how to operate a fire extinguisher would be moments before the fire starts. Since no one can predict when that will be, decisions about timing—when to train—are often based on factors that have nothing to do with learning and everything to do with administrative cost and convenience.

When do we have enough people to justify holding a class? How long has it been since the last training course was offered? How long can we go before the rules say our people have to be trained? When are the students available? The teacher? The training room? They're all legitimate questions about timing of training, but they have nothing to do with *the best time for someone to learn*.

In summary, the matter of timing is simple in concept: there is an optimum time to train, based on the relationship of the student to the material. Economics and logistics pose huge obstacles to doing that. So, instead, timing is optimized around the needs of those providing the training.

Our educational system functions the same way. Take a vacation in France and you'll wish you'd taken French class for the last two years, instead of two years back when you were in high school. Read the annual report on a company you own stock in, and you'll wish you'd taken that college accounting course last semester.

It's all perfectly understandable: that's how life works. But the mismatch between need and delivery makes for a significantly lower return on the investment in training.

THE SECOND 'T': TECHNIQUE

Adults may have different preferred styles for learning, but ultimately we all learn by doing. Whether it's how to tie our shoes, drive a stick shift, make a stock trade, put out a fire, or fill out an accident report, the process of learning reaches fruition when the student successfully performs the task.

Every sports coach understands that well, which explains why practice plays such a large role in competitive sports. In baseball it's spring training. In football it's preseason practice. Golfers know the practice tee. For team sports like basketball, football, and water polo, if it's not game day, it's practice.

Repetition is the mother of learning, and some of what goes on in practice is repetition. But there is a lot more to practice than just repetition: there's teaching and learning going on. Weaknesses are identified and worked on: "We'll keep running that play until we get it right." Improvement and skill building also go on: golfer Ben Hogan said golf was a game that must be learned on the dirt of the practice tee.

When it comes to teaching, the best teachers in sports are born innovators, constantly searching for new tools to improve their techniques. Watching game film used to be the province of the football coach's office; today, frame-at-a-time video replay is a standard tool used by coaches in every competitive sport. At their best, teaching tools allow students the sensory experience of the concept or technique being taught. Good teaching tools put more of the senses in play—touch, sight, and even sound. They can provide immediate feedback on performance, good and bad. All that makes for a training experience that is memorable, and isn't remembering the point of learning?

This trend hasn't stopped at the out-of-bounds line on the sports field. The fire service has its fire training fields; the utility industry has its version—a training field filled with utility poles. High-fidelity process simulation is part of the training for reactor operators in nuclear power generation; the manned space flight program has long relied on hard simulation of equipment to be flown in space. Perhaps the best training innovation of all time was the Link Trainer aircraft simulator, which dates back to the 1920s. Aircraft flight simulation started with what was then known as a "blue box" that replicated aircraft flight controls, developed by a musical organ manufacturer with an interest in aviation. It has since evolved into a highly sophisticated tool to teach and certify pilots.

These are but a few examples. The common thread in all of them is this: people responsible for the transfer of know-how figured out better ways to do it.

Technique plays a big role in effective training. Going back to square one, a teacher isn't limited to the conventional lecture by PowerPoint; there are many other familiar alternatives:

- Demonstration, showing what the technique looks like
- Laboratory experiments, for hands-on practice and feedback
- Reading materials, from textbooks to case studies
- Case studies and problems
- Field trips, to see the knowledge in its real-world application
- Simulation
- Class discussion
- Asking questions

Choice of technique—design—sets up how the teacher teaches.

THE THIRD 'T': TEACHER

Great teaching is not the norm, but at various points in our lives we've all had the privilege of being taught by a great teacher. The subjects taught by our best teachers varied widely—from history to physics, golf to investing. As different as the subjects and teachers might be, there are three things all great teachers have in common.

First, they all cared passionately about the subject they were teaching. Not in an overbearing way: watch them in action and you'll come away convinced that there's nothing more important going on in the world than the subject they're teaching. As a student, you can't help but get caught up in the moment.

Second, all had the ability to communicate what they knew in such a way that their students "got" what they had to say. Students learned, and what they learned stayed with them for the long haul.

Which brings us to the third characteristic that all had in common: helping students learn—causing others to know a subject—was what mattered to them. Their goal as teachers was to see to it that their students learned. Their methods varied widely; so did their personalities. But the class wasn't about them, it was about their students.

Great teachers like the physicist Richard Feynman made teaching look easy. Try doing it yourself. It isn't. If teaching were easy, most teachers you come in contact with over your lifetime would be good at teaching. In our direct exposure to teachers, from kindergarten to grad school, in sports and hobbies, military basic training and company programs, good teachers are the exception, not the rule. It exactly squares with the sentiment that there's something fundamentally wrong with much of our training...and our teachers. Most of the teaching population, including a lot of people who teach for a living, just don't teach all that well. That's not finding fault; it's simply doing the math.

Peter Drucker was of the opinion you couldn't teach someone how to teach well. Perhaps this conclusion was based on his intuitive good sense about those three common characteristics of great teachers. Passion about the subject, communicating so that students get the message, and a focus on the students are really all about the teacher's motivation.

CREATING UNDERSTANDING

Taken together, timing, technique, and teacher are the **Three Ts of Training**. Done properly, they can produce memorable—and, more importantly, effective—training. Do any of the three poorly—e.g., teach people when they aren't ready, tie a great teacher's hands with an inferior course design, or hand good course design to a poor teacher—ineffective training will surely follow.

Getting the Three Ts right is so tough a challenge, it's easy to understand why most organizations throw in the towel and just go through the motions of training, content to "tick the box" to show someone else that the training has been done.

On the other hand, consider the consequences when training is done poorly. Students falling asleep in class, is far less troubling than organizations failing to learn and improve, the same mistakes being made, and people failing to run the operation the way they should.

You don't have to be a member of senior management—or the person leading the training—to do something to improve training. The first step in the process is to become an educated consumer of training. Reading, and understanding, this chapter enables you to meet that requirement.

The second step is to recognize the situations in which you have influence and control over what goes on in the name of training. If you can pick who does the training, or when it is delivered, you have some measure of control—the ability to determine the outcome. If you can pick the trainer, what's stopping you from picking someone who has motivation and aptitude for teaching? That person might not be the one jumping up to volunteer for the assignment. For many good teachers, teaching turned out to be a hidden talent. It takes some thoughtful analysis on your part to assess who really has the right stuff—passion, focus, and the ability to communicate—to teach well.

As to timing, you can certainly resist the temptation to take the path of least resistance, scheduling training when it's convenient rather than when it's most effective. It might cost a little more or inconvenience a few people. But isn't the payoff—the return on the investment—worth the extra effort or expense?

There are the things you can't control: edicts to train in a certain way; prepackaged training materials; mandatory computer-based training; trainers sent from the home office to do the training who are poorly prepared or suited for it. But you don't have to happily accept what you've been handed. After all, you are the customer. And every customer has the right to complain when the product or service is inferior. So, raise your voice and provide the feedback. When the complaints become sufficiently loud and widespread, change can happen.

All that is a lot to ask of a busy leader. But knowledge is too important to going home alive and well at the end of the day to settle for anything less.

CHAPTER 17

MEASURING SAFETY PERFORMANCE

> None of us can see ourselves, so we have to have good information.
> If you're not getting good information, it doesn't matter how strong your desire is.
>
> —*Paul Azinger, professional golfer and Ryder Cup captain*

For over two decades we've been asking industrial leaders the world over—numbering more than one hundred thousand, from company presidents to front-line supervisors—"What are the toughest challenges you face managing safety performance?" No matter what kind of business they're in, or where in the world they're located, there is a predictable pattern to their replies. It's a list familiar to every leader on the planet: in alphabetical order, it's one that begins with attitude, awareness, behavior, compliance, complacency, communications, culture, distractions, equipment…and ends with zero: nobody gets hurt. Facing down those kinds of tough challenges is the real stuff of safety leadership.

Often lost in this discussion about safety leadership is what *isn't* on the list: measuring safety performance. Rarely does measurement show up as what leaders see as their biggest safety challenges. As to why, perhaps it's because leaders think they're already getting enough information to manage safety performance effectively. Perhaps it's because leaders think it's somebody else's job to figure out how to measure safety performance for them. Maybe it's because no one seems to know how to come up with better measures than the ones they have. Maybe there just aren't any.

Expose that reasoning to the cold light of day, the flaws become readily apparent. For openers, if the current information were really that good, no leader would ever be surprised by getting bad news about safety. In the aftermath of every problem, if nothing else, the leader would always be able to say, "I knew that was coming."

Of course, if a good leader knew there was a problem coming, they would have fixed it before, not after.

Alive and Well at the End of the Day: The Supervisor's Guide to Managing Safety in Operations, Second Edition. Paul D. Balmert.
© 2023 John Wiley & Sons, Inc. Published 2023 by John Wiley & Sons, Inc.

Perhaps it's because keeping score of injuries is too easy; the hard part is figuring out what meaningful information might be missing. The reality is that measuring safety performance presents two huge challenges for leaders. The first is making sense of the performance data they're currently getting. The second is finding better data, the kind that tells a leader what he or she really needs to know about safety performance to be able to manage performance well. All too often the numbers counted and the information reported—injuries, incidents, accident reports, inspection results, audit scores, attendance at safety meetings—provide little useful intelligence about what's really going on, and rarely provide it in time to act to correct a situation before it produces harm.

If you're a leader and you're not getting good information about the safety performance of your followers, it doesn't matter how strong your desire is to see to it that everyone goes home safe every day. As to exactly what that good information might be, the answer depends on the leader's level in the hierarchy. There is a huge difference between what an executive needs to know, and what a front-line leader requires to managing the safety performance of a crew.

We'll offer practical ideas for performance measures for each.

THE FUNCTION OF MEASUREMENT

No matter what the level or role in the operation, a common understanding of the nature, source, and use of safety information is highly beneficial to every leader. So, we'll start with the basics. Why do you measure safety performance? It's such a basic question that asking it hardly seems necessary. You might well think everyone knows the answer: "To see how well we are doing."

Yes, that statement is correct. But a deeper dive into the measurement process reveals the function of measurement to be far more complex an undertaking, a process serving what are inherently conflicting objectives. That conflict often renders information and numbers not just meaningless, but downright misleading. But to appreciate that truth about the safety numbers, you must first understand the three functions performance information performs for leaders:

- evaluating competitive performance
- trending the direction of performance
- revealing what's really going on

The first and most common function of safety measures is to judge and compare performance. While every operation and every leader strives to achieve zero injuries, no large organization has yet to achieve that goal over the long term. There are injuries to be counted; counting and comparing them is the first and most fundamental purpose of measuring. Long before workplace safety regulatory agencies began imposing injury reporting requirements on employers, the method to normalize the primary measure of safety performance—the injury frequency rate—was created to allow a comparison across operations of widely different workforce populations.

Leaders regularly compare the injury rate against their history, their goals, and the performance of others. Judgments are made: are we doing well, doing better, or doing worse? Competitive performance measures give leaders the answer.

There are those who are highly critical of this performance measure, pointing out its shortcomings. By and large, their criticisms are valid. For example, the rate does not consider injury severity: a relatively minor injury is counted the same as a life-altering injury. In the desire to improve the rate, leaders become pre-occupied with reducing the actual number of injuries counted in the rate calculation, ignoring more serious potential problems and those occurring infrequently or not resulting in harm to people. The attention on the number of injuries can cause leaders to focus their effort on the person getting hurt instead of the hazard and process that caused the harm. The most widely leveled critique of the injury frequency rate is that it is not forward looking: the rate shows what has happened, not what is going to happen in the future.

However well intended these criticisms are—no question that they are motivated by the desire to improve safety performance—they reflect a lack of understanding and appreciation of the elements that together constitute an effective safety performance measurement *system*. In that sense, those leveling this criticism are guilty of the same mistake made by those they are criticizing: focusing on only one measure of performance.

The injury frequency rate should be one measure in a rigorous safety performance measurement system. Sending followers home alive and well at the end of the day is the bottom line of safety: the number of injuries and the injury rate serves the same function as the final score at the end of the game. To ignore the result is simply foolish.

The injury frequency rate also has the benefit of universal use: it is just about impossible to find any operation or business anywhere in the world that does not calculate their injury frequency rate. It serves as a worldwide yardstick by which to compare and evaluate safety performance. A site that might be tempted to claim, "We're great at managing safety performance" need only look at how their rate compares with their peers and competitors. They might not be nearly as good as they see themselves as being.

Most industry trade groups benchmark their members' safety performance and do so using the injury frequency rate. Within an industry it's commonplace to find *more than an order of magnitude difference* between who's best and who's worst. As to what would explain why the best in the business achieve a rate ten or twenty times better than their worst competitor (and often five times better than the average performer), the leaders need only look in the mirror. As Peter Drucker put it, "Companies don't compete. Managers compete."

There is not a thing wrong with healthy competition in the journey to zero. Wanting to be the best taps into the competitive nature that is a common characteristic of business leaders the world over. For safety, the beneficiaries of that competitive spirit are those doing the work of the business.

Competitive safety performance measures also show leaders and their followers "how well we are doing" compared to "how well we've done in the past" and "how well we want to do in the future." Anyone who's ever gone on a diet knows the motivational function played by the scale. Competitive safety performance measures are not limited to the number of injuries and the frequency rate: audit scores, training performance, safety suggestions, overdue equipment inspections and behavioral safety observations can all serve as examples of what could be considered competitive safety performance measures.

Measure any competitive performance measure over time, a trend line can be calculated. Measured over time (and statistical variability accounted for) the trend will be either up, down or flat.

Reading the word trend, it's easy to jump to the conclusion we're referring to the term of art known as a leading indicator. You may have heard leaders say, "The injury rate is a lagging indicator. We need to pay attention to leading indicators." Isn't any trend line a leading indicator in the sense it reveals the direction of performance into the future?

Yes it is.

In managing safety performance, the terms leading and lagging indicators have become such a confusing and even polarizing subject that it merits its own section in this chapter on measurement. For the moment, let's consider trend data in the broad sense of leading and lagging measures.

A trend is defined as the direction of movement over time. Measures that reflect trend help get the leader get out of the current moment and observing what's happening over the passage of time. At its core, safety performance is a function of human performance. The performance humans exhibit is subject to a high degree of variability, but over time, there is always a trend: getting better, staying the same, or getting worse. The same can be said about the related processes to execute such as being fully trained, regularly inspecting equipment, maintaining equipment in good condition, keeping the work area well-organized and in good order. So training, inspections, maintaining equipment, and housekeeping lend themselves to trend analysis.

Trend data can be incredibly useful to managing safety performance. Different than competitive performance measures that show how well you've done, trend information might not predict the future, but it can alert the leader to something that might be a concern, or alternatively, to feel confident the existing programs and processes are performing the way they should.

That the safety measures would provide information as to both competitive performance and the trend of performance comes as no great revelation: that's what a leader has in mind when using the terms "leading and lagging indicators." But the leader would be much better off understanding those functions as "telling me how well we are doing compared to..." and "indicating the trend of performance for things that matter to my management of safety performance."

There is a third function the numbers have the potential to fulfill that is every bit as important as evaluating and trending. In the practice of safety leadership, this function may even be more important than either evaluating competitive performance or indicating trends: that function is to inform as to what is going on, revealing reality for better or worse.

A fundamental and inescapable truth about safety is that it is grounded in reality: hazards are objects with energy; followers work with and around hazards; things and people are never perfect. That's reality. To be able to manage safety performance well, a leader needs to *know* the current state of safety reality. If the leader doesn't know reality for what it really is, the leader is left to manage based on perception. Perception isn't reality.

In a perfect world every leader would know everything there is to know about the actual condition and state of performance for their area of responsibility. But in the real-life practice of leadership, there is always a gap in that knowledge: there is reality, and there is how much the leader knows about reality.

Safety data can provide a leader with information about current reality: for example, the equipment inspection report shows the condition of objects; the overdue inspection report shows what's not been inspected; the PPE audit shows who's wearing their personal protective equipment.

To summarize the basic principles, information coming from the measurement process can be used by the leader for one of three purposes: to compare and evaluate safety performance, to determine the trend of performance for any activity or result being measured, or to gain information as to the current state of reality. As to which purpose any measure is used, how the leader uses the information determines the function of the information: to judge competitive performance, determine the trend of performance, or to know what's really going on. That use is not mutually exclusive.

That seems obvious. The injury frequency rate serves as an example. Most often it is used as a measure of competitive performance: *How does the rate compare with our goal? With last year? How do we compare with others?* But the rate can be used to determine the trend: *What has been the rate of change over the last five years?* Similarly, the output from a behavioral safety observation program can be used to determine the trend: *What has been the percentage of safe behaviors found over the last twelve months?* The numbers can also be used to compare and evaluate performance: *How many people are actively involved in the process as observers?*

Certain numbers may serve as the means to get a sense of what's really going on. For example, the scores on graded safety audits for equipment can reveal information as to the state of their condition; the audit scores for the use of personal protective equipment can reveal information about the collective behavior of followers. The scores from audits can be tracked over time to determine the trend: are equipment conditions improving? Is PPE usage improving?

But, when choosing how to use the information, the issue leaders using the numbers must understand and fully appreciate is this: once followers know for what purpose the numbers are *primarily* being used by their leaders, the response by their followers will be to manage the numbers to produce the result their leader wants. It is what rational people working in organizations do: this behavior is predictable and universal.

Thus, the leader who fires off an angry e-mail to the team on the heels of a poor audit result may well get exactly what the leader wants to see from the next audit: a high score. Does that mean the reality about performance uncovered by the first audit has changed for the better? Perhaps. Or does it mean that the followers prepared for the next audit, knew exactly when it was coming, and made sure things looked good for the auditors? Possibly.

That is a huge difference.

This familiar phenomena is at the heart of the measurement challenge. Leaders—particularly executives who rely heavily on the safety measures to manage safety performance—must understand and recognize this aspect of their safety measurement challenge. Any measure used in any way to judge performance can't be depended on to reliably assess the trend or accurately depict reality.

Consider this the ironclad law of performance measurement. The failure to recognize this law sets up the potential for a shock to the system: a serious incident that rocks an operation where executives were sure things were safe, only to find out otherwise.

METRICS AND INFORMATION

Up to this point, we've used the term "measure" to describe the output of the process used to measure safety performance. In practice, the measurement process need not be limited to numbers, nor should it be. There is a broad range of valuable information potentially available to a leader found in the form of both metrics and information.

A metric is simply number that quantifies and counts something. As to what gets counted related to safety, usually there is lot: for example, the number of safety suggestions; the number attending a safety meeting; the number of active observers in the behavioral safety observation program; the number of deficiencies found in the monthly house-keeping inspection; the number of overdue equipment inspections; the number of near-miss incidents reported. And of course, the number of people who went home hurt or hurting.

What makes that kind of information a metric is that it is represented in the form of a number. An example of a widely used metric is the injury frequency rate: it's a number that is produced by a calculation performed following a formula that relates the number of injury cases to number of hours worked by those with the potential to be hurt.

As important as metrics are, they are only a small part of the information potentially available to leaders at every level of the organization that can be used to lead and manage safety performance. For example, the cases where people were hurt could be produced in the form of a list. So could all of the equipment overdue for inspection. A list could compare injury frequency rates by departments, sites, projects, divisions, and companies within a defined industry. That kind of information is commonly referred to as a benchmark: a benchmark is a formal measure of performance comparison.

Another common and significant source of information about safety comes in the form of reports. Reports can be formal, such as an audit report, an incident report, or an investigation report. A report can be requested or commissioned, such as an assessment from an outside expert. Often reports come to the leader informally, such as when a follower tells their leader, "We're having a problem with…"

Finally, there is what a leader personally observes as a continuous stream of information coming their way: what they see and hear from people and from all of the things—objects in the terms of the Injury Triangle—that can potentially impact safety performance. For the front-line supervisor in particular, that information is huge, and generally an accurate reflection of reality.

LEADING INDICATORS

When used to describe safety metrics, the term leading indicator is a familiar one. Familiarity is one thing; understanding is an entirely different matter. How they are put into practice more often than not serves as proof of that lack of understanding, as measures referred to as leading indicators seldom meet the technical definition of the term. Using the term leading indicator to describe a particular safety measure—or worse, relying on one—without understanding its meaning and implications isn't a good management practice. Managing safety performance already comes with more than enough challenges; no sense making a tough job even tougher by adding in confusing information.

Reading that should cause a leader to start asking questions: What defines a leading indicator? What safety measures qualify as leading indicators? What exactly does a leading indicator indicate? Any leader using or contemplating the use of leading indicators to manage safety performance must have good answers to all three questions. The answers are available, but finding them requires the leader to spend time studying what is an academic subject: statistical measurement.

The term leading indicator was imported from economic forecasting. In economics, it has long been recognized that trends do not last indefinitely. Sooner or later, there is almost always a reversal in any trend: boom turns to bust, expansion follows recession, markets go up, then down; there is regression to the mean. Because trends reverse, trend data alone is not sufficient to make a reliable forecast of the future on which to base economic decisions. Leading indicators were created to be able to detect and discern a change in trend.

In that way leading indicators supplement trend data; leading indicators are not used on a stand-alone basis.

More specifically, in economics indicators are statistics or numbers that signal or forecast: they point to something else. That suggests a relationship between an indicator and what it points to. Adding the word "leading" implies the indicator signals something that will show up somewhere else later. When the term "lagging" is used, something else showed up earlier. Leading and lagging simply describe sequential timing: before or after.

Things can be related to each other in all kinds of ways, but when it comes to designing and using leading indicators to point to something else, there are two relationships that matter: cause and effect, and correlation.

Cause-and-effect relationships are the focus of medical and pharmaceutical research. What are the causes of a specific disease? How effective is a drug under development in treating the effects of that disease? The medical journals regularly publish research findings about cause and effect, focusing on the relationship between human health and such everyday activities as drinking coffee or milk. When it comes to human anatomy, the hard science necessary to get that story straight isn't as easy as it looks. Back in the sixties a highly publicized study pointed to the health problems suffered by kids who drank milk, suggesting some kind of causal relationship between milk and poor health in children. Those researchers missed the more fundamental relationship in play: children who drank milk were far more likely to survive infancy and live to suffer the kind of health problems that show up in teenagers and adults.

That's why the control group is so valuable in proving a causal relationship. Ideally a control group exactly matches the study group, but isn't given the substance or experience being studied. A control group provides a meaningful standard of comparison.

When it comes to safety, you can begin to see the difficulty in demonstrating a causal relationship for the kinds of performance that routinely get measured. Since there is seldom a statistically valid basis for comparison, when a change is made it is very difficult to prove a direct relationship with the improvement desired. Did driver training reduce the number of accidents—or was it the change in the cell phone policy? Or something else?

The second kind of relationship between points of data is correlation. Correlation suggests a relationship, but not necessarily one of cause and effect. Correlated events happen relatively closely together, but not necessarily because of each other. Insurance actuaries have found that teen drivers have accidents at much higher rates than do adults. It's not hard to explain that relationship: inexperience, inattention, immaturity. They're kids. That's why teenagers pay higher insurance rates. Those same actuaries have also found that student driver education is correlated with safe driving. So are good grades. So those higher rates are discounted for teen drivers who take drivers' ed and get good grades.

With correlation, there isn't necessarily a cause-and-effect relationship. Getting straight As in high school doesn't cause a teenager to become a safer driver, but a report card will get a straight A student an insurance discount. The insurers see a correlation.

Confusing things are examples of correlation between completely unrelated factors, such as between stock prices and football scores or sunspot cycles. Just because a statistical relationship has been found to exist in the past doesn't mean there is a meaningful correlation. There are always coincidences. Random correlation is no basis for making important decisions about the future.

The principal benefit from an accurate leading indicator would be the ability to predict the likelihood of change in trend—with sufficient reliability to be able to make important decisions and take action based on the indicator. Were a leading indicator for safety to suggest a big problem is imminent, action could be taken to prevent that from happening.

That's the appeal of having leading indicators to manage safety. Having accurate leading indicators is a terrific idea, but creating them demands hard science, something rarely attainable in the management of safety performance. To be useful a leading metric must be accurate: one that suggests a causal relationship—*this* causes *that* to happen—or a meaningful correlation—when *this* happens, it's usually followed by *that*. In statistical science, those are known as validity and reliability. There are well established statistical methods to measure and verify validity and reliability. If those methods are not rigorously applied, the indicator is simply a theory, an opinion, or just wishful thinking.

A perfect example of what is clearly at best an opinion—not hard science—comes from one of the agencies responsible for workplace safety in the US: the Occupational Safety and Health Administration. In their effort to promote the use of what they called Leading Indicators, they published a document explaining how employers "...can use leading indicators to improve safety and health outcomes in the workplace." They begin by explaining leading indicators: "They measure events leading up to injuries, illnesses and other incidents..." and offer specific examples they see serving as leading indicators, such as the amount of time it takes to respond to a safety hazard report and the number of attendees at a monthly safety meeting.

However, there is no data provided proving the validity of either of what they claim as leading indicators, or even that attendance at safety meetings or the response time to a hazard report is highly correlated with injuries. At best these are hypotheses, and two that are easily proven false. If safety meetings are boring, why would making more people attend them have any positive effect on injuries? What's worse is what is recommended by this agency: setting a goal for improving the score reported by the leading indicator.

On the other hand, there have been statistically validated studies of safety practices—having safety meetings, turning in near-miss reports, participating in behavioral observation programs—suggesting a correlation between these events and the most closely watched lagging event, injuries. The studies can be very useful but should come with a warning label for leaders to read before use: "This is at best a statistical correlation, not a causal relationship. The correlation is only as good as the quality of the underlying data and is valid only for the organization that collected the data. Proceed at your own risk."

It's another example of the danger of mindlessly replicating a process from some other organization.

NEAR-MISS REPORTS

Without some discussion on the process of near-miss reports, and reporting, no book (or chapter) on managing safety performance would be complete. This book was never intended as a technical work for the professional safety engineer; rather, it has been written expressly for the millions of supervisors and managers whose job is to see to it that the people they supervise get the work of the business done—and go home safe. There are occasions when people doing the work have close calls, and sometimes their leaders hear about those close calls. What then? That's the practical question every leader has to answer.

Whether brought to the attention of the supervisor or not, every near miss is a Moment of High Influence, another of those seemingly everyday events that put people in a high state of readiness to be influenced. As a very senior electrical craft leadman once admitted, "At some point during their career, every electrician has a brush with electrocution." When a near miss occurs, it's bound to get someone's attention. If the injury potential was severe, the odds are good that any reasonable person would stop, reflect on what could have happened, and profit from the experience.

If the matter ended there, there wouldn't be a need to say anything more on the subject. But sometimes those close calls are witnessed by the boss or later reported to the boss. Reporting causes questions to be asked: How should a near miss be counted? What should we do about the incident? And, sometimes, what should we do to the person responsible for the incident? Those latter two questions are treated elsewhere in this book; at issue here is the question of how to count a near miss.

The simple answer: it depends on how many near-miss reports you get.

The frequency of near-miss reporting varies widely from one organization to the next. If you assume that everyone has close calls no matter what kind of work he or she does, the key word in that statement is *reporting*. As a practical matter, if near misses are seldom reported, the small number of those that are is not a useful metric.

For a leader, the practical questions to ask about near misses are: Do I think many near misses are going unreported? If one is reported, am I better off for knowing about it? If the answer to the second question is, "Yes. As a leader, I'm always better off knowing what's going on," any report of a near miss ought to be looked upon as a beneficial activity: with activities, more is always better than fewer. Moreover, if you believe that there are many near misses going unreported to management, any increase in reporting is also a good thing because you're getting more information about reality.

When a near miss occurs it creates a Moment of High Influence. When a leader first learns of a near miss, that creates a second such Moment: followers are paying attention to how the leader reacts to the news. How the leader reacts will largely determine what happens the next time there's a near miss. If there is an upside for the person immediately involved—a problem gets fixed and the person reporting it receives positive feedback for bringing it forward—more reports are likely to follow. On the other hand, if a report is met with an unfavorable reaction, further similar incidents will go unreported.

Finally, every organization has a history on the subject of near misses. If you're a leader who thinks increased reporting is a good development and even a potential leading indicator for improving safety performance, bear in mind that history dies a very slow death.

FINDING BETTER MEASURES

We live in the age of information. The amount of performance information at the collective fingertips of leaders for the business functions of production, cost, quality, schedule, and customer is huge. Increasing the amount and quality of business performance information to real time data has been a major focus of information technology projects. By comparison, information about safety performance remains relatively sparse. Given that sending people home safe is the most important business goal a leader has, should the information available to manage safety information be any different?

Of course not. What's standing in the way of doing something to get more and better information?

If you're a front-line leader or manager, you're not the one responsible for designing the measurement system for safety performance in your company; that duty falls to top management and the corporate safety function. If they don't decide to improve the process, you may well be saddled with living under a measurement system that hasn't significantly changed in 50 years.

Regardless of what those running the business do, nothing stops you from collecting and using more information about the safety performance of those for whom you are responsible. That doesn't require launching an ambitious project to create a safety database. Given the considerable expertise you already have in measuring business performance, coupled with what you now understand about the purpose, means and methods of measuring safety performance all that is needed is devoting some time and good thinking to the measurement process.

Bear in mind, "If you can't measure it, you can't manage it" and you are expected to manage safety performance.

On the other hand, you might be in the position to bring about change in how safety performance is measured at your site or for your business. As a senior leader or executive, with the power to act, what's stopping you from undertaking meaningful change in process you know is vitally important to performance?

Recognizing those two divergent viewpoints on measurement, the remainder of this chapter is divided into two sections: one for the executive—Creating A Balanced Scorecard—and a second—Measures for the Front-Line Leader.

CREATING A BALANCED SCORECARD

Executives don't need to be taught how to measure performance: it's something they do exquisitely well. The world-class measurement methods routinely used in other functions—production, product quality, cost—can readily be brought to the task of measuring, monitoring, and *improving* safety performance.

It wasn't always that way. The product quality and productivity revolution that took place in the last two decades in the twentieth century offers a useful lesson in how the measurement process can play a critical role in driving change. In the science of counting things, no one person stood taller than W. Edwards Deming.

At the height of the Great Depression, possessing a PhD in mathematical physics, Deming worked for the US Department of Agriculture. Deming's government service led to using statistical analysis processes in the 1940 US Census data. Historically, population sampling was the kind of activity in which the mathematics of probability and statistics found their application. A few years later, Deming had a better idea. With the world at war, he applied the same statistical techniques to the production of war material.

In the post-war economy, there was little commercial interest in the application of statistics to manufacturing in the US. Recognizing the situation in postwar Japan was an entirely different matter, Deming headed east.

There, Deming's statistical methods for improving manufacturing product quality were applied, and what emerged two decades later was superior manufacturing performance across a wide range of products, from electronics to automobiles. Deming's statistical methods were given much of the credit for the turnaround. In practical terms, his techniques simply gave manufacturing managers in Japan far more sophisticated performance data about their operations and products, which they used to improve both production and quality.

In the early 1980s when the US industrial economy hit the skids, Deming's methods finally got a serious audience with manufacturing leaders. What followed was a product quality and productivity revolution that carried the US economy forward for the next three decades. Although some of the disciples of Deming have tried to make them seem so, the methods weren't all that complicated. His tools of statistical analysis applied to the process of improvement—such as the histogram, the process flow chart, and the great measure of variability, standard deviation—were being taught to college sophomores in the sixties.

Armed with a more sophisticated way to look at cost, product quality, and productivity, supervisors and managers quickly perceived what needed to be done to improve. The game shifted to something line managers were very comfortable doing—altering the means and methods of production.

In the years since, these processes have continued to evolve, and are now known by names like Six Sigma and LEAN manufacturing; the basic principles are simple and powerful. The change in performance a proven fact, the lessons to be learned from the practice of measuring process improvement are substantial:

- Measurement is the sine qua non of world-class performance.
- Measurement is powerful: the right measures provide enlightenment about what is really going on.
- Measurement must be principally focused on what can be controlled: the manufacturing process as it takes place. It should not be focused on the end result: the finished product, which cannot be controlled.
- The process of collecting and sifting through data is as much a valuable part of measurement as are the findings from that analysis.
- Involving every stakeholder in the process of collecting and evaluating data isn't just wise delegation. Involvement in the measurement process builds engagement in the improvement process. The function of measuring has long since left the quality control department.
- The best operating performance metrics aren't necessarily complicated, but there are a lot of them. And the proper choice of what to measure, where to measure, and when to measure is complex, and demands thoughtful consideration.
- Relying on any one metric can be misleading and, worse, can tempt others to misrepresent the truth. No financial analyst would ever make a decision based exclusively on net income after tax.

With those measurement principles in mind, there are a significant number of options that can be considered as part of what should be looked as the "measurement improvement process." Viewed in that way, the measurement process is no different than any other work process, subject to continuous improvement.

As to where to begin, why not start with what's already in place. Create a master list of all of the measures that are used to manage safety performance: metrics, data, and information. It's best to think as expansively as possible. So, instead of simply putting "safety audits" on a list, it would be better to create an inventory of all of the types and subjects for safety audits that are performed. There are likely records of audit reports in the files: consider them data. If audits are not kept on file, or were not performed as required, consider that information, too.

Taken together, the current level of measurement intensity will begin to reveal useful information as to the current state of the measurement process. Is there enough to enable safety performance to be managed to the same level as other business functions? If not, what's missing? Are there many measures, but no way to make sense of them?

The most basic and practical defense against the tendency for performance measures to be skewed in the direction of the desired result is simply to have more measures. Increase the number of things that are counted and it becomes increasingly difficult to hide the real truth about performance. In the case of safety there are always plenty of things to measure, many of which are already being counted by someone else: safety suggestions, attendance at safety meetings, near-miss reports, audit results, inspection reports, corrective action status reports, safety rule violations. Assemble all the information that is already being reported in one place and, if nothing else, it presents a more complete view of reality. If everything but the number of reported injuries is in a tailspin, and the injury rate is reported to be improving, it might pay to be suspicious. Underreporting injuries is tempting—and easy.

Identify how each measure is currently used in the management process: is it a competitive measure, a trend measure, or a means of revealing reality. There are not any right or wrong answers as to use of any given measure; the question is simply, "How is this measure primarily used by management?" Bear in mind any measure used for competitive evaluation will compromise its reliability and validity to reveal what's really going on.

One very simple step to make the trend—the message in the data—more obvious is to employ moving averages. The moving average smoothes out any inherent variability in the data, better revealing the underlying trend it indicates. That's routinely done in the stock market, where the 50-day and 200-day moving averages play an important role in investors' decisions to buy or sell a stock. Some operations use 12-month moving averages for their injury frequency rates: it's an excellent way to see the underlying rate of change in performance.

A very effective technique to employ from time to time is to "dump the data on the table" and see what it might be telling you. The technique changes the focus from the individual element—one incident or problem—and instead looks for trends and patterns. This process benefits from a fresh set of eyes: someone with no preconceived notions as to what to expect or even look for.

Of course the first requirement is to actually have a lot to look at: metrics and information. As an example of what information might look like, picture a year's worth of incident investigation reports, sitting in a stack. For each, the event that precipitated the investigation was investigated, a report written, and preventive and corrective action determined. All well and good, but what would a reading of the entire pile yield, in terms of intelligence as to trend about problems, solutions, and execution?

Before investing time and energy attempting to develop entirely new and different metrics, consider using an existing measure as a starting point. In economics, this measurement technique is routinely used for activities that are important to understand, but difficult or impossible to directly measure. The Dow Jones Industrial Average is a composite of only 30 stocks. That average is widely followed as a representation of how the overall market for publicly traded equities—of which there are thousands—is performing. Calculating combined change is far easier for 30 stocks than for several thousand. The Conference Board publishes an Index of Leading Indicators, designed to show something that's even more important, but impossible to directly measure: where the national economy is headed. Since that can't be directly measured, an index using stand-in measures was created. The Index of Leading Indicators incorporates a series of readily available data (for example, new unemployment applications, building permits for new houses, orders for manufactured goods). The value of the index is found in the month-to- month size and potential change in the direction of the trend (which is what makes it a Leading Indicator).

Do 30 stocks accurately reflect the rest of the universe? No. Is the Dow Jones Industrial Average useful? Absolutely. Is the Index of Leading Indicators subjective? Yes. Is it useful? Absolutely.

These metrics suggest a very useful model for leaders who think that there is value in understanding what's going on in the very complex world they are responsible for managing. In measuring safety performance there may well be simple measures that perform the same function, standing in for much more complicated ones. Exactly what those easier-to-measure metrics might be is unique to each organization. Here's one potential example: safety suggestions. Most supervisors don't get all that many safety suggestions. Consider what happens when the supervisor does get one: someone who is thinking about safety sees a problem or opportunity. Then the person takes the time to present the idea to the boss, who is given a Moment of High Influence—and sometimes an opportunity to take action to correct a safety problem. If that happens, the person who turned in the suggestion sees the action, and so do others.

It turns out that embedded in the simple action of turning in a safety suggestion are a series of important actions that can eventually have a huge impact on bottom line of safety performance, the injury rate. If the number of safety suggestions submitted is steadily declining, that doesn't augur well for future safety performance. Alternatively, if the number of safety suggestions submitted is showing a dramatic increase, it's very likely a leading indicator of better performance. Something as simple as the number of safety suggestions submitted might be a very useful stand-in metric for the future direction of safety performance. If that logic is sound, it also suggests that the time it takes to respond to safety suggestions may well be a stand-in metric for management's commitment to safety.

Once assembled, all of the information can be sorted in a very useful way. Anything that can be measured can be thought of as either an *activity* or a *result*. While anything that has happened is in one sense always a result, in terms of measurement the difference between activity and result is this: activities involve discretionary action; results are the dependent variable, a product of activity (or the lack thereof). When measuring safety, results are numbers like injuries and illnesses, days lost from work, employees exposed to hazardous materials, or noise above permissible levels. Results like these are the equivalent of the golf scorecard at the end of the round; the scoreboard at the end of the game; the quality of the product after it has been manufactured. If you don't like the results, there is nothing you can do about it but try something different next time and hope for better results.

By comparison, activities are the things that are done to get the kind of results that are desired. If we are what we eat, eating is the activity; what the scale reads when we step on it is the result. When measuring safety, activities are numbers like inspections and audits performed, safety suggestions submitted, training classes taken, time spent Managing by Walking Around. If we believe more of the right kind of activity leads to better results, measuring that kind of activity—how much that activity is done or how well it is done—becomes very useful. In an ideal world, there would be proof of the cause-and-effect relationship between activity and result. In the real world of operations, that kind of scientific proof isn't usually available, but common sense often suggests a relationship. So, for example, superior industrial housekeeping might not be the probable cause of fewer injuries, but it makes sense that when the shop is neat and orderly, people are likely to be safer, if for no other reason than that there are fewer things to trip over. Attending training, performing auditing, developing procedures, participating in safety meetings, and, yes, performing housekeeping in the work area are all examples of activities. For every activity, there is some amount of discretion or choice.

Putting this practice in play, one very simple but powerful step in improving the measurement of safety performance is to measure both activities and results, and to distinguish between the two. Activities are the more important, as they are something a leader has direct influence over; results are dependent variables.

While the measurement science required for identifying, creating, and perfecting statistically significant *leading indicators* is extremely complex, the process of identifying *leading measures* isn't. That process starts with understanding what makes a measure *leading* rather than *lagging*.

Since it's impossible to measure something that hasn't yet happened, all measurements of anything are by their very nature historical. No matter what is measured—who won the big game or who was the number one pick in the college football draft, what last year's earnings were or how many new orders were received last week, who got hurt last month or who went to a safety training course—activities and results that are measured always reflect what has already happened.

That said, there are any number of ways to look at historical data. Some data reflect bottom-line performance: who won the game, how much money the business earned, how many people got hurt. Like the scoreboard at the end of the game, numbers like those are the most important because they measure success or failure.

But those numbers reflect only what happens at the end of the process: the game, the year, or the month. They aren't the only numbers that are collected and reported. There are normally also numbers that measure things that happen in the process but don't show up at the bottom line: new players signed on to the team, new orders received from customers, the number of people who attended training courses. Those measures might track activities that are important to someone, like sales calls for the sales manager. They might reflect intermediate results that are factors ultimately important to success, like the value of new orders or the number of customer complaints.

What differentiates those numbers on the basis of leading and lagging is the matter of timing: when they show up in the process. Some of those activities and results show up early in the process, and some show up later on. The last thing that always shows up is the score at the end of the game. The prime objective is to win the game, but when the final score is posted, it's too late to change the outcome. So football coaches track who is recruited and signed to play on the team, how many attend off-season conditioning programs, what the players' strength measurements are, and how those measurements change with time in the weight room. Every football coach knows waiting around to see the final score is no way to coach the team to a winning record.

Measures that track the data that show up early in the process are *leading measures* and the ones that show up later are *lagging measures*. It's a simple distinction and applies to all kinds of activities. In the business world the stock market is normally considered a leading measure of economic performance. Stock prices go up or down in anticipation of future earnings. So when the stock market declines significantly, falling stock prices may signal recession. And it may not: hence the old saying that "the stock market has predicted 17 of the last nine recessions." Movement of a leading measure may or may not signal change in the lagging measures, but it's a sign. By comparison, change in employment levels is a classic lagging measure of economic activity. Normally firms hire only after they are convinced there is permanent work, and lay people off when the work those people do is no longer needed. Changes in economic activity often show up in employment figures last.

If a process has been stable for a long time but it begins undergoing some sort of change—an indication of the change would be evident in a well-designed and functioning leading measure. A useful leading measure may reveal what will happen later on, and what will show up in the lagging measures. Applying this simple logic to safety, it's easy to detect the leading measures: sending people to training courses, conducting more audits, writing new procedures, launching a program, spending more time managing by walking around. Those kinds of measures can change rather quickly. So can the results of those activities: audit scores, test grades, measures of compliance, and safe or unsafe behavior.

Change is far slower to show up in other measures of safety performance such as an increase in safety suggestions, voluntary reporting of near misses, improvement in the quality of accident investigations, or a reduction in the number of injuries.

As to where to draw the line between leading and lagging—faster or slower—that's entirely a matter of choice. What's more important is to recognize that not all numbers will change with equal speed. When it comes to predicting the future—and doing something about it before it's too late—smart leaders simply will know where to look first: to the leading measures.

Any measure of safety performance can be seen as either an activity or a result. Any measure can also be seen as either leading or lagging. These two observations about measures can be combined, forming a matrix that represents a **Balanced Scorecard** for Safety. The balanced scorecard (Figure 17.1) enables you to view all the principal measures of safety performance at one time, thereby reducing the temptation to focus on the injury frequency rate to the exclusion of other meaningful measures of performance.

THE BALANCED SCORECARD

	Leading	Lagging
Activities	✓ Training ✓ Inspection ✓ Housekeeping ✓ MBWA	✓ Safety suggestions ✓ Near-miss reporting ✓ Using specific skills ✓ Safety observations
Results	✓ Inspection results ✓ Audit scores ✓ Test scores ✓ Certification	✓ Injuries ✓ Accidents ✓ Frequency rates ✓ Violations

Figure 17.1   The Balanced Scorecard provides visibility to a number of leading and lagging measures of both activities and results.

There are other significant benefits that come about from the process of creating a balanced scorecard. First, developing one means aggregating all the measures of safety performance from all the unrelated places they are kept: the safety training records system, monthly equipment inspection reports, safety suggestion system, quarterly safety audits, and injury records. Finding all that out is in itself instructive. It should reinforce the notion that improving the measurement of safety performance doesn't necessarily require you to create new data.

Another benefit of developing the scorecard comes from deciding where each safety measure belongs. The matrix provides four potential choices for any measured data: it can represent a leading activity, lagging activity, leading result, or lagging result. What are the differences? More importantly, what's the point served by each of the choices?

Here's the logic: activities—whether leading or lagging—are those things over which a leader can exercise some degree of direct influence. A leader can call safety meetings, send people to training, commission safety audits, and schedule equipment inspections. However, a leader can't determine what the results of those activities will be: what people actually learn in the training course, the number of deficiencies found during the safety audit, the condition of the equipment when it is inspected. That's the distinction between an activity and a result.

Some activities happen much faster than others, in part because many activities depend on other important factors that can't easily be directed or influenced. For example, a manager can normally require everyone in the department to attend a safety training class…that's a leading activity. If the course teaches safety leadership practices, how well those practices are learned by the students attending would represent the leading result. That dependent variable, learning, is a function of several independent variables: the aptitude and interest of the participants, the effectiveness of course design, the skill of the teacher. A test given at the end of any course is not a bad way to measure how well those factors combine to form training performance. But the primary objective in sending people to training is not for them to pass the test. If the material taught in the class is how to more effectively lead others to work safely, the ultimate measure of success will be found in the future, in the injury rate.

That injury rate is a lagging result. It is a dependent variable, one that can't be dictated by the leader. Many safety metrics, particularly the ones that get the most management attention, are lagging results. They are necessary, but relying exclusively on lagging results is analogous to trying to "inspect quality into the product after it's been made," as Dr. Deming put it.

But injury frequency rates are a lagging result, the last thing to move either up or down. Looking at them is like looking in the rearview mirror. Applying moving averages elsewhere in the balanced scorecard has even more potential benefit: if a highly variable leading activity is showing underlying decay, a moving average will reveal that trend. So, for example, leading activities like turning in safety suggestions or leading results like audit scores or equipment failures can benefit from the application of moving averages.

What is missing from the example of measurement as it applies to training are the subsequent leadership actions taken by those attending the course. The presumption is that if leaders attend the course and learn and practice the leadership skills taught, their followers will be more inclined to work safely. The more those skills are practiced, the more likely safety performance is to improve. That makes practicing leadership skills an activity, albeit a lagging one.

Attending safety leadership training starts out looking like an ordinary event that might show up on the scorecard as a leading activity. Delve into the process, understand training as an investment of time and resources, and chase the return on that investment in the direction of bottom-line results, and all kinds of things become clear. There are four places to look for measures: participation in the class, aptitude on a test, the practice of better leadership skills out on the job, and, finally, improvement in the bottom line of safety, the injury rates. In that order, those four reflect a leading activity, leading result, lagging activity, and lagging result.

The process of constructing the Balanced Scorecard requires the one-time investment of thinking about those relationships. Sorting the independent variables— what the leader actually influences—from the dependent variables reveals those relationships and suggests where to look to see how performance is changing before it shows up in the "finished product." It helps focus on the things the leader can actually influence. Thinking about timing—when changes should or might start showing up in the process—provides a better understanding of functioning of the underlying processes (for example, what is really required to increase the number of safety suggestions or near-miss reports). In the process of collecting and reporting the measures, a more complete picture about performance emerges.

Supervisor training is one small example of how the balanced scorecard can help. Apply that same good thinking to the many other measures of safety performance that are already being recorded, and you can begin to see the power locked up in the balanced scorecard. Using it is a great way to understand the measures you have and to use all of them to get a more complete picture of what's going on at every stage of the process of managing safety performance.

Safety performance metrics don't have to be cast in stone. There really is no reason why performance measures can't evolve over time or be changed to suit the circumstances. If your organization is experiencing high turnover, and you have a lot of new, inexperienced people, performance measures may be aimed at producing data telling you how well the new guys are learning to do their jobs safely. That data might include a leading result, average score on qualification testing, and a lagging result, the injury frequency rate for short-service employees. If you're making major changes in safety policies and procedures, you might create a new measure of a leading result, the rate of compliance on new and revised safety procedures.

MEASUREMENT FOR THE FRONT-LINE SUPERVISOR

We now return to the front-line supervisor: the subtitle of this book is *The Supervisor's Guide to Managing Safety in Operations*. If you're a front-line leader and thought as you've read this chapter, "This is not my problem" you would not be wrong. The measurement challenge is principally one faced by senior leaders and executives. They live a long way away from where the work takes place, are responsible for managing large numbers of followers, need information to evaluate safety performance and to see reality for what it really is. As far as seeing reality for what it is, as a leader at the front line, Managing by Walking Around and convincing your followers to tell you about their near misses will take care of most of that need for you.

Moreover, at your level, injuries requiring treatment more than routine first aid are rare. The statistics suggest if you supervise a crew of twenty, and their safety performance is average, you will see an injury requiring medical treatment once every few years. That might lead you to conclude you really don't have a measurement problem: everyone is working safely

But, in practice you do have a measurement problem. It's just a different problem than that faced by your top management. You need to have a good sense as to the direction the safety performance of your followers is headed. That is your sphere of influence and control. For that you need information about their trend: getting better, getting worse, or staying the same. The problem is that you have no injuries to count to tell you that.

That is a nice problem to have, but it is still a problem. Your team's safety performance could be living on borrowed time: they are headed for a problem—incident or injury—but haven't gotten there yet. If you knew that was the trend, you could do something to change it. On the other hand, if they've never been safer, don't change a thing.

This is the point where you might think, "That's what we have leading indicators for. All I have to do is to check them, and they'll tell me how my followers are doing." There are two flaws with that line of thinking.

The first you will understand from reading the earlier part of the chapter where leading indicators were examined. In practice, very few so-called leading indicators for safety stand the test of statistical correlation: they're usually just someone's opinion as to what might precede a change in the trend of safety performance. They might be right, and they might not be.

Appreciating what you need at your level to determine trend exposes a second flaw found in even the best of measurement systems: leading indicators reflect the performance of *all* followers—not the performance of *your* followers. As such, you may have no idea how well your followers are actually performing. That flaw deserves further explanation.

A supervisor regularly reports safety performance information for their area and followers. At the end of the shift and at end of the week, a safety report is filed. It might be part of a shift record or production report. Should someone be hurt or property damaged, an event report would be filed. When people are trained, it's recorded in the human resource record system. Suggestions are turned into the safety suggestion system; near-misses are turned into the incident reporting system. Behavioral safety observations are turned into the behavior-based management data base.

Raw data the individual leader turns in is always combined with the data from all the other leaders. Reports reflect everyone's performance. That's fine for the senior leader who is responsible for the performance of all followers. That is not so for the leaders who report to the executive. For each of them, the information is very likely to be misleading because the assumption is made that every follower is the same, and every leader's followers are the same.

Picture a front-line leader supervising a crew of twenty in an operation employing two thousand: that leader's crew represents 1% of the total population. Can that leader safely assume their followers and their safety performance is identical to every other crew in the operation?

Of course not. But that is the presumption made whenever a collective performance report is published: every crew is achieving the same level of safety performance. Across a company, variability is commonplace between divisions, sites, and projects. Why would it be any different at the individual leader level?

Your followers could be performing better than their peers—or worse. If that's the case, the performance feedback you give them based on the leading indicators is likely to be counterproductive. You tell them, "We're doing poorly" when they are not; you tell them "We're doing great" when they are headed for a problem.

Ironically, it might be better to have no data.

So, even if you are convinced your company has world class safety metrics, you now understand that even under the best of systems, the performance measures for your followers—who represent the sphere of your influence and control—are mixed in with the measures for all followers of all other leaders in the data base.

That's why it's essential for you to have trend measures indicating the future direction of safety performance of your followers, your equipment, and your processes and their products.

That is your measurement challenge. But what leader has time to create a complex system to collect and analyze that kind of information?

You don't have that luxury, and you absolutely should not even think about taking on the technical challenge of creating statistically valid leading indicators.

You don't have to, because you have available all the information you need to get a good sense of the trend of the safety performance trend for your followers **Early Warning Indicators**. You just need to appreciate what that information is and know how to organize it to put it to that use.

Let's call the kind of information you use to evaluate the safety performance trend for your followers as **Early Warning Indicators**.

Early Warning Indicators are defined as information used by a leader to forecast the future direction of their followers' safety performance. Your Early Warning Indicators perform the same function as do leading indicators for the senior leaders: a sense of the direction of safety performance: getting better, staying the same, or getting worse. Their leading indicators need to be statistically validated to assure they are reliable indicators. Yours need only to be based on your judgment, experience and knowledge of your followers.

But your Early Warning Indicators must focus exclusively on what is under your control and influence as the supervisor. That differentiates an Early Warning Indicator from a leading indicator: you can't rely on some other leader's performance trend to tell you yours.

As to the information serving as your Early Warning Indictors, you're not required to meet the same standard as does a formal leading indicator. Your information doesn't even need to be numbers. In the words of Albert Einstein, "Not everything that can be counted counts, and not everything that counts can be counted." You are looking for information: metrics and things like conversations and appearance.

What information might serve as an Early Warning Indicator for you (see Figure 17.2)?

Start with the easy stuff: measures that are already collected, processed and readily available to you—and that specifically (and only) applies to your crew. For example:

- Overtime rate
- Turnover rate
- Safety suggestions
- Near-misses
- Incident reports
- Minor injuries
- Audit reports
- Inspection reports

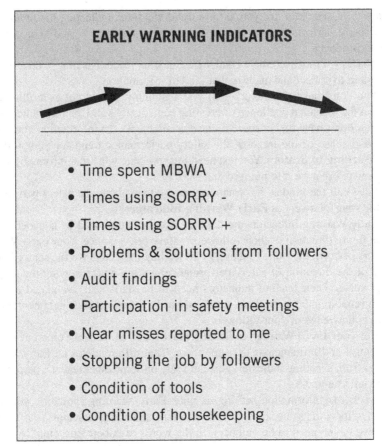

EARLY WARNING INDICATORS

- Time spent MBWA
- Times using SORRY -
- Times using SORRY +
- Problems & solutions from followers
- Audit findings
- Participation in safety meetings
- Near misses reported to me
- Stopping the job by followers
- Condition of tools
- Condition of housekeeping

Figure 17.2 Examples of potential Early Warning Indicators forecasting the future direction of the safety performance of followers.

Next, look for activities you do and results you see first-hand. For example:

- Your time spent doing MBWA
- House-keeping conditions
- Your observation of your followers' behavior

Since Early Warning Indictors are not limited to hard numbers, qualitative information can also be very valuable. For example:

- Frequency and nature of questions asked by followers
- Degree and quality of participation in safety meetings
- Peers intervening when they see someone not working safely
- Reaction to being coached from the supervisor and/or peer
- Followers exercising Stop Work Authority
- What followers are talking about

In other words, you can use any information you want as your Early Warning Indicator—as long as you're reasonably confident the information is telling you what you need to know. An Early Warning Indicator doesn't have to be perfect to be useful, but it does need to be reliable. If it's not, relying on it can be harmful, not helpful.

Will this practice add to your workload?

Yes. To analyze the safety performance trends requires additional effort above and goes beyond what's found in the definition of your duties and responsibilities. Initially, you'll have to invest good thinking into developing your Early Warning Indicators. Asking the question at the end of every workday, "Is the safety performance of my followers getting better, staying the same, or getting worse?" will have to become part of your management practice. Otherwise there is no point to having these Indicators.

Your Early Warning Indicators have one huge advantage over all other safety performance measures: they're yours! That makes them immune to becoming a measure of performance used by your leaders to judge and evaluate your performance. Freed from that, the indicators can tell you what you're looking for.

And you will have to listen to what your Indicators are telling you.

THE LAST WORD

Whether you are an executive, manager or front-line supervisor, you are now armed with practical ideas about measuring safety performance in ways that can put you on the road to better measures. What you don't have is "the answer" in the form of one great metric that will tell all. Instead what you have is a way to think about the measurement process, recognition of the need for more and better information, a framework for understanding and evaluating the functions of your safety performance measures, and specific avenues to create the kind of safety measures that provide the information you need to be able to manage safety performances well.

As Paul Azinger so aptly put it, "If you're not getting good information, it doesn't matter how strong your desire is."

CHAPTER 18

MANAGING SAFETY DILEMMAS

On the horns of the dilemma.

—*George Santayana*

This book takes on the tough safety challenges standing in the way of sending every follower home, alive and well at the end of the day. Many challenges are of the type every leader understands perfectly: lack of knowledge, experience, poor training, and ineffective safety meetings fit that description. There are challenges a leader recognizes, knows, but does not completely understand: complacency, compliance, hazard recognition are examples of challenges that are far more complex than first meets the eye. There are challenges that often go unrecognized: executives need unvarnished information as to safety reality that is kept separate and distinct from data to determine bottom line safety results.

Then there are the dilemmas leaders routinely deal with in managing safety performance. When facing a situation that fits this category, a leader is more likely to see it as a failure on their part rather than what it is: a unique type of safety challenge that defies a conventional solution.

Tracing the word back to its Greek origin, *dilemma* comes from the words meaning "two assumptions." In a dilemma, present are two conditions that are both true and in opposition to each other. It is the dual nature of the dilemma—and the tension it produces—that led to Santayana's description of being "on the horns." It's the perfect metaphor for these tough challenges. Every leader in operations regularly comes face to face with a series of real-world dilemmas —the Accountability Dilemma, the Risk Dilemma, the Investigation Dilemma, the System Dilemma, the Middle Dilemma, the Leader Dilemma—with horns perfectly capable of impaling the best-intentioned (see Figure 18.1).

As a leader facing one of these dilemmas, in the heat of battle you can't call a time-out to study and analyze the particular nature of the challenge you're dealing with. In practice, you might not even recognize you're in the crossfire of conflicting conditions that are the byproduct of the dilemma. What you witness in the moment is

Alive and Well at the End of the Day: The Supervisor's Guide to Managing Safety in Operations, Second Edition. Paul D. Balmert.

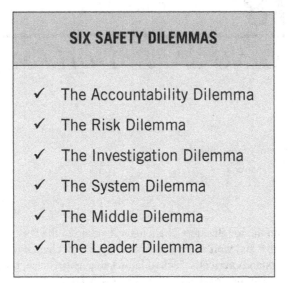

Figure 18.1   Six safety dilemmas.

a problem for which there appears to be no good solution. The effect on you is totally predictable: enormous frustration.

If you need examples to understand the challenge, consider these.

- Despite your best efforts, your followers are not working safely, not following the rules, and still getting hurt. You're taking the heat from your manager and your performance rating takes the hit. Welcome to the Accountability Dilemma.
- The work is planned, the hazards identified and managed, but that doesn't guarantee nobody gets hurt. Zero harm seems impossible, That's the jumping off point for the Risk Dilemma.
- When investigating a failure, nobody owns up to the truth about their role in what went wrong. Things just happened; it's nobody's fault. That's the Investigation Dilemma.
- You learn an injury happened to someone who broke a rule that nobody follows. Is enforcement just another management failure—or the System Dilemma?
- The leaders you work for don't want to hear about the problems you face in managing safety performance, particularly the serious ones that are above your pay grade to solve. Are you a poor communicator—or is it the Middle Dilemma?
- In your performance appraisal, you're coached to become a "stronger and more visible leader on safety." But your crew's safety performance is superb. The Leader's Dilemma is lurking beneath the surface.

Find yourself on the horns of any one of these dilemmas—sooner or later you are bound to—you don't need a warning label to know the horns are sharp. But owing to the nature of the dilemma, seldom does the leader recognize they're on the horns of one, particularly in the moment. Quite the opposite: the leader is inclined to see the situation as the result of their shortcomings as a leader. All they can think to do is to bear down even harder—and wind up going home with a headache.

Dilemmas really are tough challenges. Don't think for a moment managing them will ever be easy: the inherent nature of a dilemma—two assumptions or conditions equally true and in opposition to each other—means there are no simple solutions. That's the nature of the beast.

Managing safety dilemmas is first and foremost a matter of being able to recognize and understand the source of the challenge. Knowing what's really going on gives you the best tactical advantage possible to deal with the conflict.

1. THE ACCOUNTABILITY DILEMMA

> As head coach, you feel a lot more responsibility with a lot less control.
> It can be frustrating. I put the plan out there and rely on the players to
> execute it.
>
> —*Football coach and former quarterback Danny White*

When the players don't execute, guess who gets fired?

The coach draws up the play, and the players execute. In competitive sports, the performance and the job security of the coaching staff is left entirely to the on-field execution of the players. Most of those players aren't anywhere near as competent as their coaches were when they played the game. Often the coach has little say as to who's on the roster. But when the players don't execute, it's the coach who takes the hit.

That same thing happens with leaders at every level, and in all walks of life: there is always a gap between what a leader is held accountable for and what the leader can control or influence, which you now know to call Organization Power. This is the Accountability Dilemma. It's the reality about being the leader that's a source of significant frustration for line supervisors and managers. The easiest way out of this dilemma—or so it appears—is simply to ignore it.

In the case of safety, ignoring this dilemma means hoping nobody gets hurt. Hope is not a method.

The Accountability Dilemma is usually the first tough challenge a leader experiences after they're promoted into management, when they go from being accountable for what *they* do to being accountable for what *others* do. It is the definition of a supervisor: the person responsible for the work of others.

Frustrating as it can be—it is—the Accountability Dilemma comes with the job.

MANAGING THE DILEMMA

Just because there's no way to make the Accountability Dilemma go away doesn't mean the situation is hopeless or unmanageable. There are ways to lessen the stress this dilemma produces on you. If you're a supervisor and on the horns of this dilemma, here are four ideas that can help.

IDEA 1: RECOGNIZE THAT YOU ARE ON THE HORNS OF A DILEMMA

Pretending that the dilemma doesn't really exist will only make things worse. Your boss will hold you accountable if someone working for you gets hurt. Don't expect anything different. You'll never succeed in controlling what those working for you choose to do, so don't waste valuable energy trying to "get control." Leadership doesn't work that way.

There will always be a gap between what every leader controls and what they are held accountable for. The challenge only gets worse as you move up the chain of command: if you think it's tough being a front-line supervisor with a crew scattered all over the operation, consider the CEO, who's accountable for what everyone in the company does, everywhere and every time.

That doesn't mean that nothing can be done to narrow the gap between your accountability and your organization power. The smart thing is to know where to spend your energy: that has nothing to do with your accountability and very little to do with the principle of control.

IDEA 2: CALL THE DILEMMA BY ITS PROPER NAME

It's the Accountability Dilemma. *Naming* it is the first step in *taming* it.

Naming something provides all sorts of useful benefits, starting with providing a degree of objectivity. The name gives you perspective: this isn't really about you. Every boss faces the Accountability Dilemma—simply because they are accountable. Naming the dilemma creates the opportunity to put the dilemma up on the examining table, probe around for symptoms, and reach a diagnosis: "Yes: this is a classic case of the Accountability Dilemma."

With a name, you know what to look for, and how to treat the symptoms. Note that we did not say, "cure the problem."

IDEA 3: LEAD BETTER

The best way to deal with the Accountability Dilemma is to make the gap between accountability and control as small as possible. Of course, no leader controls their followers. As you know, the correct way to think about what many leaders call *control* is to think of it as *influence*. There are things that every supervisor can control, but they are *things*—not *people*.

General Dwight Eisenhower described the process this way: "Leadership is the ability to get someone else to do what you want done, because he wants to do it." The best leaders do that, and do it so well that the gap between control and accountability is very small.

So lead better. Look at your own behavior and performance as the leader; measure it and determine what needs to be improved. Then set out to do better at what you can control—your own behavior as the leader.

As to what to do and how to do it, the answers are found in the chapters of this book. Follow the advice, and your gap will be lessened.

IDEA 4: LOOK TO YOUR FOLLOWERS

If the leader doesn't have control over the actions of those they supervise, who does?

The answer, of course, is the individual followers themselves. The very best at safety leadership have been successful in convincing their followers they are the ones ultimately accountable for the results. When it comes to safety, those doing the work have the most to gain by working safely, and the most to lose when they do not.

Those doing the work have control over what they do. For followers, there is no dilemma. If you can make your followers feel just as accountable for their safety as you are as their leader, you can live with the Accountability Dilemma.

2. THE RISK DILEMMA

> Many of us are struggling with the concept of risk; how much is too much, how much is an inherent part of what we do?
>
> —Wayne Hale, Space Shuttle Operations Manager

Manned space flight is among the riskiest of human endeavors. The images of the space shuttles *Challenger* and *Columbia* serve as graphic reminders of the potential consequences when risk management fails.

The need to successfully manage risk isn't limited to the scientists and engineers of NASA. The need is universal. "How much risk is too much? How much is just part of what we do for a living?" If you're the supervisor of a crew in operations that does the work of the business, you've had to have asked yourself the same question.

Every day you spend time managing risk; so much so the process of managing risk seems routine: writing and approving work permits, performing hazard assessments, developing and communicating risk control plans and procedures, handing out personal protective equipment. It's easy for leaders to fall victim to "ticking the boxes" so followers can start work. It's easy for followers to take the same approach.

Suffer the misfortune of one of those headline making failures, the processes to manage risks and their consequences get put under a microscope in a root cause failure analysis. Everyone is reminded of the critical importance of these processes; a heightened sense of awareness follows.

Hale's statement about risk was prompted by the Columbia accident in 2003. But organizations are populated by humans, and with passage of time that keen awareness dissipates, giving way to complacency. Case in point: seventeen years had passed since the last shuttle accident.

Risk is an inherent part of what people contend with when they're at work; in some way *managing* risk is a big part of everyone's job. Risk often goes by unnoticed: it's a normal—and acceptable—part of the job. That's also the way many regard the methods used to manage risk.

Most supervisors and managers are too busy managing risk to take the time to reflect on what makes it a challenging undertaking. The process of managing risk involves the usual suspects—people, equipment, materials, and physical environment—none of which are ever easy to manage. Perhaps that's why a thoughtful leader like a NASA Mission Director would use the word *struggle* to describe the effort.

There is more to the Risk Dilemma than first meets the eye: wrapped up in the process of managing risk are both a dilemma and a conundrum.

THE DILEMMA

What makes for a dilemma is for the two vital conditions to be beyond dispute—and in complete contradiction. As with a battery, the greater the difference between the poles the greater the voltage—and the bigger the shock when you make contact. The Risk Dilemma fits the image perfectly.

Risk is defined as hazard times probability. The definition implies two elements in the risk equation: what can produce the harm— hazard—and how likely it is to harm—probability.

Everyone has plenty of practical, everyday life experience in applying this definition of risk. It might rain: should I carry an umbrella on my morning walk? The highway conditions have become treacherous: should I pull off the road and wait until they improve? I'm moving to a new home: should I buy flood insurance? If there weren't any costs involved in managing risks like those, we'd always carry an umbrella, wait out the storm, and buy insurance for anything that could go wrong. But providing for contingencies comes with a price: energy, time, money.

That's where the Risk Dilemma starts.

To protect people from the risk of getting hurt on the job, an assessment is conducted. In theory, the process is simple: determine the hazards, i.e., figure out what might go wrong that could get someone hurt and then take appropriate steps to prevent it from happening. That's what is done at the start of the shift in the pre-job hazard assessment and safe work plan.

If you examine the process very closely, you'll begin to appreciate that this hazard-management process was never intended to cover every hazard imaginable, nor to reduce the probability of every hazard to zero. If you took that approach, no one would ever leave the pre-job safety briefing. In practice a risk management process considers the potential seriousness of the consequence. If it's loss of life, there's one answer. If it's a paper cut, there's a different answer.

Assigning someone to work off the ground might require the use of fall protection—unless that person is working on a stairway landing with a permanent handrail. In that situation, it's assumed the handrail eliminates the risk of falling. But handrails can fail; it happened to someone leaning on one and when it broke, he fell 10 feet and died.

Does that mean you should require fall protection for anyone going up and down a stairway and holding on to a handrail? Of course not. It's not practical to do so—it's inconvenient and expensive, and the probability of a handrail failing are very low.

But not zero; so should you do a risk assessment for using the stairs?

In actual practice, the risk assessment process in operations goes something like this: first, the hazards most likely to lead to injury are identified. Next, *reasonable* precautions are taken for the hazards that are *reasonably* likely and *relatively* serious. There is no standard by which to evaluate *reasonably* and *relatively*. Left off the list is a long list of hazards considered unlikely or effects inconsequential. For hazards put on the list, measures are taken to reduce probability or severity. Their risk isn't zero; they're acceptable.

That's the first horn of the Risk Dilemma: the list of things that can go wrong on any given job—hazards—is large; the probability of any given hazard occurring is not zero. If it were zero, it would not be a risk.

Acceptable risk means stuff can happen. What happens when it does create the second horn of the Risk Dilemma?

THE TRUTH ABOUT CONSEQUENCES

In the abstract risk is a simple concept. But the moment the bad thing that you didn't want to happen does, risk is no longer an abstraction: it has become an event, and events have consequences. Consequences change the face of risk: a follower you know and care about gets hurt, and you're the one responsible.

Viewed after the fact, the acceptable risk taken is now unacceptable. There's no solace in the fact that the event happened but once in ten thousand times. When you tell people the goal is a workplace free from injury, you really mean that. After a life altering injury, nobody would ever think to say, "Sometimes those things just happen. As long as they don't happen too often, I guess that's OK."

The Risk Dilemma boils down to this: it is impossible to eliminate every risk. Yet when the unwanted event occurs, nobody wants to live with the consequences. Leaders want it both ways: take risk but have no events. Leaders don't get to have it both ways.

That's the Risk Dilemma.

THE RISK CONUNDRUM

Wrapped up in the risk dilemma is the Risk Conundrum. A conundrum is an intricate and difficult problem, which perfectly defines the second challenge of managing risk: if you define risk as hazard times probability, how do you know what all the hazards are? Even if you think you do, how do you accurately estimate risk—the probability each potential hazard occurs?

Ponder the answers to those two Darn Good Questions, you'll come to appreciate the intricate and difficult problem—the conundrum—you face in managing risk.

But this is something you do every day.

THE HAZARDS

Identifying hazards doesn't seem difficult. It's done all the time: somebody makes up a list. All that takes is some experience on the subject—and a little time. Most of what goes wrong in operations isn't all that unpredictable.

Suppose you're the one responsible for coming up with the list—performing the hazard identification. You'd start by thinking about hazards. How many different ways could someone get hurt doing a specific job, like re-lamping the warehouse? Spend 10 minutes, you'd create a list; spend all day, you'd create a much longer list.

In operations, risk management normally focuses on the most *likely* hazards— not *all* potential hazards. Otherwise, the risk assessment would never end, and the job would never start!

So, for openers, just because a way someone could get hurt isn't on your short list of hazards doesn't mean it couldn't happen. Here's an illustration: a two-member survey crew measuring the height of a structure. One goes to the top with a steel tape measure, while the other one waits at the bottom to catch the tape when it's lowered. A wind gust catches the tape, sweeping it in the direction of a nearby power line. It makes contact, shocking the one holding the reel.

Think that could never happen? It did. Think if it did, it wouldn't be serious? It was. Think it was on the list of hazards as part of the pre-job assessment? No way.

Here's a second illustration: You know that water contacting electrical equipment is a hazard and, because there is a lot of electrical equipment and a lot of sources of water, the probability of something like this happening in any kind of operation isn't really low. It can happen when a mine floods or a facility endures a hurricane—and maybe even a leaking roof during a thunderstorm.

What about the water coming from a backed-up toilet? Think that could ever happen?

That's exactly what happened in a power plant. The result: water and raw sewage contaminated the control system. What happened next wasn't all that unpredictable: the procedures for drying out the equipment weren't precisely followed. When the unit started backing up, control systems began to short out. The control room operators didn't imagine that their controllers might be failing because of that problem, so they just ignored the alarms. That chain of events ultimately led to a boiler explosion, and damage in the range of 500 million dollars.

Fortunately nobody was hurt.

If you were performing the risk assessment, which of these hazards would you have put on your list? Probably not the hazard created by a stopped-up toilet, or the tape contacting the power lines way over there. If someone brought the possibly up as you were making your list, you'd be inclined to say, "That'll never happen." How many leaders would take the time to think about the possibility of something like that happening, and if they did, decide the odds were high enough to worry about it?

In practice, managing risk requires making tough choices about which hazards are worth the time and attention to manage. How likely does a hazard need to be to make it worth the effort? One in 10? One in 100? One in a thousand? One in a million?

In practice, these decisions are based on experience. Factored into the hazard identification process are hazards from events that have happened; often left out are the ones that have never happened.

Until they do.

MANAGING THE DILEMMA — AND THE CONUNDRUM

What do you do about the Risk Dilemma? Which horn do you want to give in on—elimination of all risks, or living with the consequences of the occasional failure?

How about neither? What about dealing with the Risk Conundrum? Is it enough for a leader to continue to apply the conventional wisdom about the hazards to pay attention to and ones to ignore? That's no way to lead.

Managing risk is tough. If the dilemma and the conundrum were easy, they would have been solved long ago. It's tempting to ignore them: they consume time and energy, and they'll never be completely solved. But they keep showing up in high-profile tragedies, and many ones that don't make headlines. You can't wish them away.

If you're serious about the Case for Safety, you've got to do something. Perhaps the best you can do is to reduce the odds that you'll see their effects show up on your watch. Here are some simple ideas to help do that.

IDEA 1: AVOID BEING TRAPPED BY ABSOLUTES

You'll never eliminate all risk from every job. Wiser to think in practical terms than in absolutes. Instead of trying to manage every risk—including the risk of being hit on the head by a falling piece of aircraft debris—the better tactic is to go after reducing the *next risk*: the hazard most likely to show up that isn't being managed well. That's the process known as continuous improvement. Your goal should always be to move in the direction of less risk, not more. On that point, every small step helps.

IDEA 2: THINK BETTER ABOUT HAZARDS, RISKS, AND HOW TO MITIGATE CONSEQUENCES

In operations it's easy to focus on the hazards with huge consequences—the scary things—and miss the potential for failure in everyday hazards simply because they're so familiar. If you're a supervisor, once you realize you can't manage every risk, it becomes easy to just stay in the comfort zone—in this case, performing the work processes the way they've always been done. You won't get in trouble doing that, and you'll always pass the safety audit.

Doing that can also mean turning a blind eye to what everyone knows is a significant risk because it's a hazard that is not easily or readily solvable. If it were, someone else would have fixed the problem a long time ago.

There are always better ideas—we just don't always come up with them. At least not before somebody gets hurt. After that happens, collectively we come up with all kinds of solutions that nobody ever thought possible.

Don't wait for a serious injury to do that. Turn loose the creative juices. Get everyone involved in the solution process. Use the practical techniques to develop better solutions described in the chapter, "Managing Safety Suggestions."

3. THE INVESTIGATION DILEMMA

Fix the problem, not the blame.

—*Catherine Pulsifer*

Every leader who's ever sat in an investigation meeting, trying to figure out what really went wrong, knows all about the Investigation Dilemma.

There is a noble purpose to every investigation: find out what went wrong so the problem can be fixed and prevented from happening again. All well and good, but consider this: in learning everything there is to know about the problem, what are the odds there won't be human fingerprints on the causes?

The reality is that a complete and thorough investigation will find out not just *what* went wrong, but *who* did wrong. Unless it was an Act of God, it can't be any other way. That human factor is what creates the dilemma. Everyone knows that going into the investigation, so no one wants their fingerprints to be found on those causes.

What that produces is totally predictable and perfectly understandable: there's a strong incentive to make sure the root cause of the problem is found in something else—or someone else. That's why "defective objects", "the management system" and "the safety culture" get more than their fair share of being named the root cause.

It happens all the time, in failures small and large, and the behavior isn't limited to those with "hands on tools." Fear of consequences drives behavior at all functions and levels. A classic example of the phenomena took place in the course of the investigation of the space shuttle Challenger whose investigation team was chartered by no less than the President of the United States. In the middle of the investigation, a member of the Presidential Commission, Nobel Prize–winning physicist Richard Feynman, broke the story that an O-ring failure was to blame for the *Challenger* accident.

As to how one of the independent members of the investigation team came to find out about the technical details of this failure—shrinkage of the seal in cold temperatures—Feynman later revealed the story in his autobiography. Someone on the inside of the space agency—an engineer who knew the real story on what went wrong and possessed the test data to prove it—left a package with all the details on his doorstep one morning—anonymously.

As important as finding the root cause of a headline making failure costing seven lives was, that space scientist wasn't willing to tell all face to face to the blue-ribbon panel of investigators. The story is decades old; the behavior and its motivation behind it is not; no one likes to own up to their failures. It can show up in any investigation.

It would be nice to think that an investigation is "an unbiased search for the truth" but sometimes the truth can be just too hot to handle. Yes, statements made by leaders like "fix the problem, not the blame," and following a root cause investigation methodology can help discover what really went wrong, but as long as there can be consequences when people do something wrong—as there must always be—the Investigation Dilemma will be alive and well.

Now you understand why you dread going to investigation meetings: when you're the leader and it's your followers who are involved, you are on the horns of a dilemma.

MANAGING THE INVESTIGATION DILEMMA

Enough about the problem: what can you do to manage it? Here are four practical ideas you can apply to the Investigation Dilemma.

IDEA 1: RECOGNIZE THIS DILEMMA WHEN YOU FACE IT

As with the Accountability Dilemma, recognizing the beast is the first step in the process. Everybody involved knows what's really going on, so why not just admit it right from the start? Pretending there isn't any tension in an accident investigation isn't fooling anyone.

So, for openers, don't fool yourself: as the leader you are on the horns of this dilemma. Consider putting the truth out to those involved in the investigation. You could do that in the form of a Stump Speech— "I know this isn't easy, because nobody wants to find out they played a role in a failure, but it's essential that we find out the truth so nothing like this happens again."

IDEA 2: INDEPENDENCE HELPS

Big failures—space shuttle accidents, aircraft accidents, transportation accidents, major process incidents—are investigated by a team of independent investigators with no direct stake in the findings. That's no guarantee: they still depend on people telling them the truth about what they know. Some facts speak for themselves, but it's often far easier for an outsider to ask the tough—and important—questions to everyone.

IDEA 3: MAINTAIN YOUR PERSPECTIVE

Getting in trouble pales in comparison to what everyone ought to fear the most: if the truth isn't found, the conditions that caused the failure won't be fixed. That means that sooner or later the same thing will happen again. You don't want to live with that on your conscience.

That's the Case for Safety.

IDEA 4: BE PREPARED

Bear in mind the real root cause of this dilemma: if people didn't fear the consequences of being in the wrong, this dilemma wouldn't exist in the first place.

But then again, if people didn't fear the consequences of being in the wrong, the world would be a far less safe place to work. That fear keeps a lot of people from doing the wrong thing and going to the lengths necessary to do the right thing. That's a good thing.

Looked at that way, you really want and need the Investigation Dilemma.

4. THE SYSTEM DILEMMA

> No snowflake in an avalanche ever feels responsible.
>
> *—Voltaire*

It is very likely that no single individual in the twentieth century had a greater impact on product quality than did W. Edwards Deming. To a practicing statistician, as Dr. Deming was, the world of performance takes on the shape of a bell curve. The best and the worst—and everything clumped in the middle—aren't all that different, in a statistical sense. They're all products of the same "system."

And what was the system Deming was talking about?

A system is the complex relationships between related components. The idea comes from the natural world—think ecosystems—and from recognizing that even small changes can rattle up and down the world we live in. In terms of manufacturing, the system is all the factors in play to make the product and to serve the customer: raw materials, production equipment, methods and processes, and people and all that their presence in the system implies.

Deming argued if you wanted better results, you had to change the system that was producing those results. Quit blaming the people for making poor-quality products; focus on changing the system that produced those products.

It was an argument that carried the day. Those engaged in making things— cars, consumer electronics, chemicals, paints, parts—started applying statistical methods to their production techniques. They would change the process, move the mean, reduce variability, and tighten up the distribution curve. The results were nothing short of astounding. Product quality improved, and so did cost and productivity, and, ultimately, profitability.

Deming was a genius, his impact profound.

WHERE'S THE DILEMMA?

With a success story like this, you're probably beginning to wonder where there could possibly be a dilemma. Or what any of this has to do with managing safety performance. That's because you've only heard half the story—and the better half at that. Remember, every good dilemma has two conditions that are equally valid...and totally at odds.

Dr. Deming was right. There are systems, and those systems are often a significant factor in determining results. The logical way to change performance is to change the system.

A key component in these systems are the humans who design, build, operate, maintain, diagnose and correct the systems they are an integral part of. Unlike every other component, humans come fully equipped with the ability to choose how they act. This is not an inconsequential difference. A human can decide whether to follow a procedure, fix a problem, and even whether to fix a system that is badly underperforming.

All of which creates the System Dilemma: human performance is a function of the system people operate. But, by the choices they make, humans determine the performance of any system they operate.

Yes, we humans are creatures of the system, but not unwilling creatures. So, how much human behavior is a product of the system and how much is a product of how humans choose to behave?

That's the System Dilemma.

THE IMPLICATIONS

Suppose you're unhappy with safety performance and the behavior of people in the system who produce that performance. Before you follow Deming's logic to "fix the system" and wind up in a place you might not want to be, you need to consider this: if the system determines behavior, the system—not the individuals in it—bears the responsibility for poor performance.

As a leader, you can't hold the system accountable. Are you willing to let the system excuse the behavior of individuals? Doing that means people are not responsible for their behavior: the system is responsible for their behavior.

In reality, the system must belong to somebody; management hast to be responsible for the system that you're unhappy with. So, exactly who are "they"?

This is the System Dilemma in full relief. Yes, there is a system that determines performance, and, yes, individuals in the system determine their behavior. Both statements are true; both statements are in opposition.

When a leader ignores the first, results probably won't change for the better. But ignoring the second becomes an invitation for irresponsible behavior, by followers who the leader would expect to know the difference between right and wrong.

By way of illustration, picture an accountant working for a publicly traded company who falsifies financial records to make the business performance look better. They're thinking about keeping management happy, earning a bonus, doing what's necessary to keep their job, at the expense of the company's shareholders.

The same logic holds for safety: picture a leader who takes shortcuts, signs waivers, ignores the warning signs of unsafe equipment, and doesn't stop unsafe jobs. They're thinking about keeping the customer happy, earning a bonus, doing what's necessary to keep their job, at the expense of the potential harm to their followers.

Neither examples are hypothetical: the former took place at Enron, the latter on a drilling rig named Deepwater Horizon.

MANAGING THE SYSTEM DILEMMA

Unlike the other dilemmas, in one respect the System Dilemma has a simple solution. No matter what the system, nobody *has* to do something they know isn't right. But don't think for a moment *simple* means *easy*!

IDEA 1: RECOGNIZE THE DILEMMA

This dilemma often comes with a warning label: "The System." Whenever you see that term of art applied to problem—cause, effect, or solution—it's time to pay close attention: the dilemma may be in play.

But it may not: sometimes pointing the finger at the system is the easy way out, because pointing the finger at humans who are not behaving well is not.

IDEA 2: THINK LIKE A PARENT

Setting aside the leader/follower model for a moment, as a reasonable adult you know someone does not have to go along if he or she really doesn't want to, just because "everyone else is doing it." Parents don't buy that rationale when it's offered by their kids; why should leaders think differently?

IDEA 3: THINK CRITICALLY

There are situations and circumstances where the system should appropriately be considered as the problem or the solution. That's Deming's view. There are situations and circumstances where the individual's choice of behavior should appropriately be considered as the problem or the solution.

You may be thinking there is a simple way to decide between the two: there is not. This challenge demands critical thinking. Do not make the mistake of falling for simplistic thinking like, "Fix the system, not the person."

To manage this dilemma, you can't sidestep investing time answering the question, "When do I fix the system and when do I work on improving the performance of my follower?"

Finally, do not overlook the role you play as the leader in making it easier for your followers to do the right thing, and making it harder for them not to. Thinking about the answer to that question puts you in business of addressing both horns of the System Dilemma.

5. THE MIDDLE DILEMMA

> Today, many managers, while working harder than ever, are experiencing a greater sense of disempowerment.
>
> —Bob DuBrul

Credit Bob DuBrul and Dr. Barry Oshry with inventing the term *Middle Dilemma* some 30 years ago. As systems consultants, Barry and Bob had an uncomplicated way of looking at organizations: no matter what the nature of the organization, there were only three roles that mattered: tops, members, and middles.

Their principal interest was in the role played by those in the middle, who link members with tops. Bob's and Barry's appreciation of the middle role came from their work with a wide range of middles: waiters, camp counselors, church pastors, and, yes, supervisors and managers in the world of industry. You can see by the list that *top* and *member* describe a wide variety of roles that aren't limited to our traditional view of the levels found on an organization chart.

The model may seem simple (the best ones always are) but that doesn't mean it isn't useful, or fascinating—particularly as it explains the difficulties faced by those functioning in the middle of a system.

A waiter—who links the customer and the kitchen—provides the perfect illustration of the difficulty of life in the middle. The waiter takes the order, the kitchen staff prepares the food, and the waiter serves the meal. When the food doesn't meet expectations, guess who bears the brunt of the criticism from the customer? It's not the chef, who's back in the kitchen, far removed from that particular heat. The waiter has no control, and often very little influence, over what goes on in the kitchen. The kitchen staff is completely insulated from the customer and seldom has to deal face to face with its own failures. That's a duty left to the waiter, who, by the way, is working for tips. How much of a tip do you think an unhappy customer leaves for the waiter? Or, for that matter, when was the last time you picked up a best-selling book written by a waiter? Like every good middle, waiters labor in obscurity.

You're beginning to see that DuBrul and Oshry were on to something. Middles play a vital role in all kinds of organizations and operations, but it's one that leaves them feeling powerless, all too often caught in the crossfire between the two parts of the system they connect. It's a frustration that most of us have experienced at one time or another. But when it comes to managing safety performance this situation can be more than frustrating: it can be downright dangerous.

THE MIDDLE DILEMMA AND MANAGING SAFETY

By now, if you play any sort of middle role in your organization, you've probably jumped ahead to the connection between the Middle Dilemma and managing safety performance. Just like our waiter, those working in the middle in the world of operations feel the brunt of the decisions and actions made by those at the top of the organization. They hear about it from the members they manage; sometimes they even get to witness firsthand what happens when things go wrong.

While the function of a middle is, by definition, linking members and tops, in practice that linking function serves as a layer of insulation between those two levels. That insulation makes life easier at the top, sparing them of those seemingly unimportant details, petty problems and the gripes members always seem to have. It can also mean that important information gets sidelined instead of passed up or down. When that happens, the tops don't know what middles know about what's really going on. That's not a good thing. It's not necessarily the fault of those at the top. Those in middle roles—with job titles like front-line supervisor, area superintendent, and process engineer—live out where the real work of the organization takes place. They are familiar with all the details that matter about safety performance, such as the real qualifications of those performing the work, the true condition of the equipment, how well policies and procedures are being followed, and what the real performance data looks like.

Said another way, middles know reality; those at the top may not.

ABOUT REALITY

If those at the top always acted as if they understood this—and were hungry for the truth—there wouldn't be a safety version of the Middle Dilemma. Even though they should, the tops don't always want to hear the details of organization reality in *their* organization: it's messy, confusing, and sometimes contradictory to what they would prefer to think is reality. That puts the middles on the horns of the dilemma. A middle can tell management all—and get branded as "alarmist" and "obstructionist" or simply not doing their job well. Or a middle can drive reality underground—and be viewed as a "can-do" type.

Taking the path of least resistance is always easier—at least until there's a serious problem. When that happens, the tops claim they are shocked to learn what's really going on. "How could you let that situation exist?" they ask the middle. There's never a good answer to that question.

Know the feeling? If you do, you've got plenty of company.

Back to the Challenger investigation and the O-ring that failed due to cold temperature, when a late-night conference call between NASA's space shuttle management team and its rocket propulsion contractor team serves to illustrate the problem as well as any ever documented. The contractor offered their customer sound, engineering-based reasons why the shuttle should not be launched at a temperature of less than 53 degrees. Frustrated by the implications, a senior NASA official blurted out, "When do you want me to launch, next April?" Whether intended or not, remarks like that, coming from tops, can have a significant effect. It certainly did in this case. The contractor's senior executive on the call then instructed his technical experts to put on their "management hat" to appreciate the customer must be served. Thereupon, he and his team decided the hard science wasn't nearly as important as being a "can do" supplier.

MANAGING THE DILEMMA

Now that you understand what's going on, what do you do about it? The Middle Dilemma is a serious problem, acutely so for safety. Most of the important action happens between the middle and the top. The two principal factors to be dealt with are the restricted flow of important information coming up the chain of command from the middles and the insulation of the top from organization reality that produces.

IDEA 1: DON'T HIDE THE TRUTH

In the aftermath of the Challenger, NASA recognized that a big part of its culture problem stemmed from the fact that top management had become cut off from science and engineering. Read that as "removed from the hard facts of reality."

Let's face the truth: middles have created much of this problem themselves. Their desire to make their operation look good to those at the top causes them to act in very predictable ways: cleaning up the place before the big visit, showing the boss the newest and best parts of the operation instead of the oldest and worst, putting the best spin on the hard data about condition and performance. Given two versions of reality, middles are naturally inclined to report the best case.

If you're a middle, try tilting the information direction of the center. Disclose some of the bad with the good. Air a little bit of your dirty linen. In the short term you might not look as good, but in the long term you're probably better off with the tops understanding your reality.

IDEA 2: PRESENT REALITY BETTER

Yale professor Edward Tufte built a successful career teaching how to present reality...better. A master of the chart, he taught that the conventional means of communication (read that as PowerPoint slides) do anything but present reality well. Tufte said "the PowerPoint templates (ready-made designs) usually weaken verbal and spatial reasoning, and almost always corrupt statistical analysis."

Countless technical and management presentations to the tops serve as evidence that most of middles don't explain things very well. Yes, they know reality well. But all too often that reality gets lost in a flood of acronyms and confusing data presented in a rapid succession of PowerPoint slides.

Take a lesson from those in the advertising business: keep the message simple, and don't be reluctant to repeat it. If all else fails, try communicating the old-fashioned way. Talk to people. When Louis Gerstner became president of IBM he sent a powerful message to his organization when he asked the presenters to turn off the projector, and said, "Let's just talk about your business."

IDEA 3: THERE IS POWER IN NETWORKING WITH PEERS

Middle management has information, and that information represents a huge form of power. Oshry wrote, "Middle space is a potentially powerful space. Individual middles are more knowledgeable about the total system, they are able to provide more consistent information to others, they are better able to provide guidance and direction."

The military long ago recognized the power of information: on the battlefield, good intelligence can spell the difference between victory and defeat. The problem with having good information in the middle of the organization is that it tends to be parceled up in small packages, each held by a different member of middle management.

Picture that information fully networked and integrated, and you can begin to see the power that is locked up in the middle. Back at the restaurant, if the waiters were to compare notes and figure out how much business was being lost by poor-quality food prepared by the kitchen, and share that information with the chef and the owner of the restaurant, you can bet there would be changes.

As noted by Oshry and DuBrul, the problem is that the system operates to pull middles apart—actually up or down the hierarchy—and they often see their peers as either unimportant—or worse, as the competition. As DuBrul observed, the management function of information sharing becomes the boss's job—not the middles. Instead, were middle managers to realize how much power they could collectively wield by laterally pooling their information to create useful intelligence, it might be impossible for the tops to be able to claim they weren't in the loop.

IDEA 4: REMIND THOSE AT THE TOP WHAT'S AT STAKE

Leading has always been a tough duty, but in today's culture, leaders live and lead in a politically sensitive world. Followers don't just take offense, they report the offense, knowing it will put the leader on the defensive. Recognizing that potential, leaders choose their words carefully so as to offend no one. The truth becomes collateral damage in the effort to keep the peace.

That might be a good way to quiet the hysteria and avoid offending. It also can lull people into ignoring a serious problem. Sometimes blunt language is what is needed to inject a healthy dose of reality into the situation. The CEO at Alcoa, after hearing a report into a fatal accident involving a 20-year-old employee, turned to his senior management team and said, "We killed him." He knew full well that it would not make for a pleasant conversation.

Bottom line, proceed with caution. But safety is that important, so by all means proceed.

6. THE LEADER DILEMMA

> You can accomplish anything in life, provided that you don't mind who gets the credit.
>
> —*Harry Truman*

The Leader Dilemma is an entirely different animal from the other dilemmas. While it's very troubling, it is first and foremost about leadership at the top of the organization.

That comes as good news, as most supervisors and line managers work and live a long way from the corporate suite. But don't get too comfortable. If you're a leader at any level, and you want to send every member of your team home safe at the end of every day, there is something very important in the Leader Dilemma that you need to understand. That's because this dilemma gets to the heart of what it takes to be the leader—and what kinds of leaders produce the best results.

The dilemma starts with understanding leadership prototypes. Which makes the best business leader?

Type A: The high-profile visionary, a superb communicator, a tough and unrelenting driver of change.

Type B: The self-effacing, limelight-avoiding tactician who's inclined to focus on continuous improvement and sweat the details.

Now, these are an oversimplification of individual leadership styles, designed to make a point. No leader exactly matches either description, and there are plenty of good leaders who don't match either.

On the other hand, if you've been working for any big organization very long you've probably seen enough of both types of leader behavior to know them when you see them—and you probably have an opinion.

So which type of leader do you think is better?

The modern theory of management teaches leadership techniques right out of the Type A description: figure out the vision, sell it to the organization, set about orchestrating the grand strategy to accomplish it. Stay relentlessly on task; leave the details to others to figure out.

That's what good leaders are supposed to do, right? There are plenty of books on leadership written by or about a Type A leader to support the case. On the other hand, those Type As aren't always the easiest bosses to work for, which may well determine the choice about who makes the better leader.

Better than holding a popularity contest to determine the answer would be to collect the data as to how organizations perform under the two types and let results speak for themselves. Best-selling author Jim Collins did exactly that, as detailed in his book, *Good to Great*. The answer stunned everyone, even the author himself.

Collins' research data proved the pattern of leader behavior described as Type B was single most significant factor in achieving business results Collins described as "great." Companies in the great group were led by a succession of leaders he described as "self-effacing, quiet, reserved, even shy—these leaders are a paradoxical blend of personal humility and professional will."

Certainly these were leaders who communicate and possess a clear sense of direction. But they were consummate no-names: passionate about getting sustained results, but more than happy to give others the credit and having no need to be in the limelight by doing something like writing a book about themselves.

THE DILEMMA

You're probably wondering: where's the dilemma? It's there, but you have to look for it.

If the best leaders at transforming organizations are the quiet, self-effacing types who are passionate about improving the business, how do they get noticed in the first place? That leadership style is bound to have typified their career, meaning they were always the types who quietly go about their business, getting great results. They probably gave a great deal of credit to everyone but themselves: their subordinates, their boss, and even good fortune. If it weren't for Collins' bottom–up research approach—he started with performance and worked up to leadership and these leaders—nobody would even know who these Type B leaders were.

Alternatively, those Type A high profile leaders have all the notoriety: they're the ones giving interviews, making speeches, writing books. Collins refers to them as "I-centric" leaders; his research clearly indicates they aren't nearly as effective as the Type B leaders. When a leader becomes the center of the organization's universe—as the Type A can become—that leadership style is not likely to produce world-beating results.

The most effective leadership style is very likely to leave the best leaders under recognized and under appreciated by the organizations they work for. The style far more likely to get noticed and recognized is exactly the style that isn't likely to get the best performance. That's the Leadership Dilemma.

If that's the case for the top executives, it may well be the case for you, too. Meaning that if you're really that good at leading, you may never earn fame or fortune. The Leader Dilemma applies to individual leaders, and to organizations in the way they evaluate leaders—present and future.

MANAGING THE DILEMMA

If there was ever a doubt, Collins' research proves that leadership is the difference that makes the difference to business performance. If it's true for business performance, there is no reason to think safety performance would be any different. That's the take-home message for all who share the goal of leading people to work safely.

Collins makes the point that leaders who transform performance are fanatically driven to produce sustained results for the good of the organization no matter what it takes, and no matter who gets the credit. In managing safety performance, that motivation is called the Case for Safety.

It's that type of leadership that's needed at every level of the organization: leaders hold the safety of those who work for them in their hands. Make the followers the focus of the effort: it's all about followers going home safe to pursue all the really important things in their lives.

Keep your eye on that target, and don't think you have to be a Type A leader to get great results.

Because you don't!

THE LAST WORD ABOUT "BEING ON THE HORNS"

Pretending a dilemma doesn't exist will only make things worse. Appreciating that two opposing conditions can be simultaneously true and in play helps the leader deal with the inevitable conflicts. Yes, the leader is responsible for safety, and no, the leader isn't in control of everything going on. Yes, the leader needs to know the truth, and no, the leader won't always be given the truth. Yes, there is a system, but, at the same time, everyone gets to decide for themselves. Yes, leaders need to lead, and no, the leader can't be the focal point of your followers.

Often these dilemmas show up in Moments of High Influence: after something goes wrong or when a new leader assumes command. That means the leadership activity that follows—words and actions—is watched closely by the rest of the outfit. Given that's the case, managing safety dilemmas well demands that a leader be on their best game.

CHAPTER 19

LEADING FROM THE MIDDLE

He who walks in the middle of the road—gets hit from both sides.

—*George Shultz*

Front-line supervisors, superintendents, and department managers all belong to the wide swath of the organization known as middle management. No matter what the business or operation, middle managers play the vital role of linking those at the top, who set policy and determine direction, with those who perform the work. Strong leadership in the middle—from the front-line supervisor on up—is absolutely essential to see that everyone goes home alive and well at the end of every day.

Leading from the middle isn't limited to leading followers who directly report to the leader. There are times when the target for a leader's influence are peers, customers, and sometimes even the boss. Doing that requires leadership. While the principles of leadership don't change based on the audience, when the "followers" are peers, customers, and superiors there are obvious practical considerations to be dealt with that makes leading tough.

While the contribution of middle management in leading their followers to work safely has been illuminated in the proceeding chapters, as have best leadership practices, organization power has not. Understanding power in business organizations is essential to wielding influence. The fact is that middle managers seldom understand what kind of power they actually have, where it comes from, and how their power can be effectively used in situations where the target of leadership isn't their direct followers.

Friday 4:48 p.m. On The Customer's Site

Everything on the project had been going perfectly—until a last-minute snag: the new sign showed up late. Very late on Friday afternoon. Upon learning that, the owner went off on the contractor: "Everyone in town knows the grand opening is at 8 a.m. tomorrow, and I can't open without my sign. You promised me you'd be done on time, and you need to fix this!"

Alive and Well at the End of the Day: The Supervisor's Guide to Managing Safety in Operations, Second Edition. Paul D. Balmert.
© 2023 John Wiley & Sons, Inc. Published 2023 by John Wiley & Sons, Inc.

The contractor tried to reason with the irate customer: "Look I understand but the sign was just delivered by the supplier. We need to build a scaffold to hang the sign. It's gonna take a cherry picker to hoist it up. And I need to call out my electrician to connect up the power. It's almost 5 p.m., and the earliest we could possibly get all that done would be noon Monday."

The customer wasn't buying any of that. "I've done enough construction work myself to know you don't have to do it that way. Get some strong backs out here, use a couple of those step ladders off your truck to get the sign up. It isn't that heavy. Anybody can tie in the electrical. I'll hook up the power myself, if that's what it takes.

There's a newspaper reporter coming to the grand reopening in the morning, and my sign has to be in the photo they'll publish. I'm paying good money for the coverage.

My sign better be up by 8 a.m. —or else!"

The customer didn't hear what the contractor muttered "You better not be standing under the sign if it'd dropped when my crew is carrying it up those ladders."

UPWARD LEADING

The Leadership Model graphically demonstrates two fundamental principles of leadership: leaders have followers, and lead followers by words and actions. The model applies no matter what the setting—business, civic, military, social—or the level of the leader—CEO or front-line supervisor. We normally picture the directional arrow of leadership pointing down the chain of command, but there are occasions when the directional arrow of leadership runs in exactly the opposite direction: it's up the chain of command (Figure 19.1).

In those situations, the principles and process of leadership do not change—it's still a matter of using words and taking action to influence "followers" —but in this

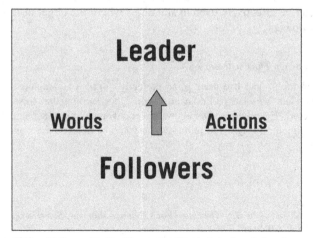

Figure 19.1   At times, leadership can be directed up the chain of command.

UPWARD LEADING

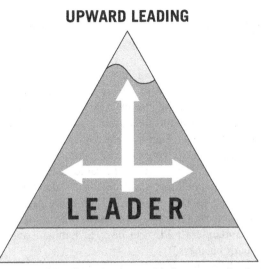

LEADER

Figure 19.2   When Upward Leading, the target of influence can be the boss, the peers or the customer.

setting the practice of leadership feels entirely different. That's because there are dynamics in the process that are different: the customer can take their business else-where, peers don't have to follow, and the boss is always still the boss. Acknowledging that reality, we'll call these situations where the target of influence is the boss, customer or peer **Upward Leading** (Figure 19.2).

When you find yourself in a situation that calls for you to engage in Upward Leading, it's normal to see your "follower" as holding all the power, leaving you feeling powerless. In our scenario, on that Friday afternoon when the customer insists "sign up—or else" that's exactly how the contractor feels.

FEELING POWERLESS

In the conventional view of the organization pyramid, the higher up, the more impor-tant and powerful the leader. That model makes the CEO the most important and powerful member of any organization. In the opening chapter, we upended that way of thinking about who's most important. Measured in economic terms, those who do the work of the business to create value for the customer are the most important. Few CEOs would disagree. Their value makes their front-line leader a critically important member of middle management. To those doing the work, their front-line supervisor is the face of the company. When it comes to managing safety performance, nobody plays a more important role than do the leaders at the front line.

But "playing an important role" and "being the face of the company" is not to be confused with power. **Organization Power** is defined as the ability to make what you want to happen.

In an organization, each level has its own inherent and defined power, and the power vested in the leaders is evident to the rest of the organization. The top is seen as having the most power. For example, top executives have the power to set the goals: goal-setting determines what everyone else is to do. Top executives make financial decisions and commitments that generate the wherewithal to create value by investing in the plant, equipment, raw materials, technology, and research, and underwriting the sales effort to create customers for the business. They define strategy and set policy.

In short, top executives run the business.

At the other end of the organization hierarchy, on a day-to-day basis, the power held by those who do the work to create value often goes unrecognized—until they're gone from the scene. Lose a key resource at the value line because they been injured, retired, or hired by the competition, suddenly comes the realization: "Wow, they were carrying a huge load for the rest of us." If the entire value line were to collectively decide to engage in a work stoppage, no value is created. At least not until staff and management members are told, "Put on your hardhats and gloves and get to work, or we'll be out of business."

Thoughtful middle managers can appreciate the power held by both top management and the value line and conclude that there is little power left in the middle for them. Rarely do middle managers appreciate the considerable power that they could wield, both down and up the chain of command. And because they don't appreciate the power they have, they don't use their power.

Actually, the real story is slightly—but significantly—different. Because a middle manager doesn't appreciate the power they have, that power is not used in a constructive way. But it is used, albeit often unwittingly. When it comes to safety, a front-line supervisor has a huge amount of power. Not understanding that power causes its misuse, a factor found at the heart of incidents and injuries, including some of the biggest headline-making events.

To put their significant Organization Power to use in a constructive way, every leader in the middle—supervisor and manager—must understand the nature of organization power. That requires undoing one of the biggest misconceptions that has grown up around the practice of management: the difference between control and influence.

CONTROL VERSUS INFLUENCE

The terms control and influence were defined earlier, in Chapter 11, "Understanding What Went Wrong." The definition was explained in the context of finding good solutions to problems, but the significance of control and influence extend far beyond fixing problems. The two terms define and determine a leader's power, no matter what the level of the leader.

In the context of managing safety performance, the terms *influence* and *control* are familiar to every leader. "We need an energy control plan." "We need to get control of the situation." "Nothing we've done has had any influence on safety performance." "We've hired one of the most influential consultants in the business" to help improve performance.

The words might seem to be interchangeable—and leaders do use them that way—but by placing influence and control side by side and comparing them, the difference begins to become apparent. The simple question, "How much control—or influence—does a supervisor have over the safety performance of their followers?" begins to make it clear, there is a difference.

Even if you're not sure you can articulate that difference, every leader knows there is one, and knows the difference between control and influence is real, and not inconsequential.

Giving operational definitions to both terms helps illuminate their difference. For that, let's return to those offered earlier in the book: control is defined as the ability to determine the outcome. "You can control the temperature of the room by setting the thermostat at 72 degrees."

Influence is an entirely different proposition: influence is the ability to produce an effect without the use of force or a direct command. The Greek philosopher Plato was greatly influenced by his teacher, Socrates. When General Eisenhower described leadership as "the art of getting someone else to do something you want done, because he wants to do it" there's no doubt he had influence in mind.

Understood in that light, control is absolute—and simple. Control is a guarantee: if you want the room cooler, turn the thermostat down. It's called a temperature *control* device for a very good reason. By the same logic, you can't *influence* a change in the temperature in the room. Influence is a process that applies to people. Short of locking people up in jail, affecting the behavior of others is a matter of influence, not control.

Properly defined, part of the confusion over control and influence is now resolved. But not all. A second question gets to the heart of what is greatly misunderstood: "As a leader, which is the better of the two forms of power to have and use: control or influence?

The vast majority of leaders will instinctively reply, "Influence." Their reasoning is simple. Followers don't like to be controlled; controlling leaders are disliked by their followers. When followers are controlled by their leaders, they don't make decisions for themselves or for the right reasons. In the long run, performance will be better and safer when followers buy-in to things, so they'll make the right decision whether or not the leader is there.

Besides, influence is easier!

The argument makes perfect sense. It's also completely wrong. As to why, the explanation is simple: control means "the ability to determine the outcome" so with control, a leader gets exactly what the leader wants, and nothing they do not. Whereas, with influence, there's never certainty because the outcome is determined by someone other than the leader.

That logic is irrefutable. It explains why, in the case of managing hazards, engineering controls are preferable to a safety procedure that requires compliant behavior. With control, the result is guaranteed; no need to depend on anyone else to follow the procedure. Or not.

You might quickly object, "People don't like to be controlled." But people can't be controlled. It's the attempt to control that people object to. In practice a leader can only influence followers—even when the leader is standing right next to the person, constantly reminding them what to do. That's not control, it's stifling influence. It's still up to the person to decide whether to do what the leader tells them to.

So, for the leader, control—the ability to determine the outcome—is always preferable to influence—where the determination of the outcome is in the hands of someone else. Taken together, control and influence determine power, the ability to make things happen the way a leader wants them to. Therefore, a leader's power is the sum of their control and influence.

You now have a proper understanding of control, influence, and organization power.

CRITICAL SAFETY FACTORS

Organization Power is the sum of control and influence. Exercising control and influence defines the practice of safety leadership. The exercise of influence is found in leadership activities such as persuading, convincing, selling, encouraging, motivating—and sometimes, if that's what it takes, talking about consequences. Safety leadership practices found in *Alive and Well* such as Stump Speeches, MBWA, SORRY, and holding a follower accountable following the 5 Ss are examples of exercising the power of influence.

As to the power of control, you might think a leader has no control, particularly a front-line leader, who feels like their power is limited to influence, and most days with very little of that. That is one more example of the misconception leaders have about power. Every leader has the ability to control their behavior, do they not?

For example, how well a leader puts into practice the leadership activities described above—Stump Speeches, MBWA, SORRY, 5 Ss—is entirely under their control as the leader. But it is equally correct that the choice to be influenced by those practices is under the control of their followers.

The power to control is not an abstract concept. As to a leader's power to control—the superior form of power—do you think it begins and ends with a leader's practice of leadership? Could there be anything else of significance under a leader's control that matters to managing safety performance?

Finding the answer begins by examining a list of factors that have proven critically important to safety. Consider these **Critical Safety Factors**: identified by looking into headline-making safety failures to understand what went wrong (Figure 19.3). These factors aren't the deep, underlying root causes like "management system failure" or "normalization of deviation" or "failure to learn." These factors are simply the immediate (or proximate) failure causing the event: the last thing to fail that, had it not, the event would not have occurred. In terms of the Fundamental Questions, they are the answer to what, how, and who—not why.

As to the factors found on the list, there are tools, equipment, methods, procedures, plan, schedule, training, qualifications, staffing, assignment. None of which should come as any surprise. Over decades, this familiar set of factors has caused immense harm to people at work.

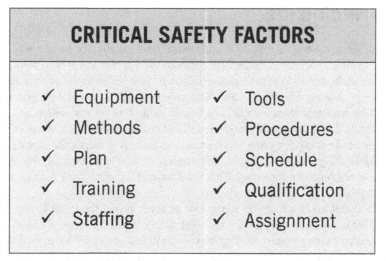

CRITICAL SAFETY FACTORS

✓ Equipment   ✓ Tools
✓ Methods   ✓ Procedures
✓ Plan   ✓ Schedule
✓ Training   ✓ Qualification
✓ Staffing   ✓ Assignment

Figure 19.3   Ten Critical Safety Factors account for the proximate cause of many high-profile safety events.

As one example, two Critical Safety Factors were responsible for Union Carbide's Bhopal tragedy: defective equipment—a tank external cooling system, scrubber, piping and flare that all failed—and procedure—a Standard Operating Procedure limiting the volume of product in storage tanks wasn't complied with. Critical Safety Factors do not explain why: in this case, why was the equipment in such poor condition or why procedures were not followed by operating staff. Those explain what is commonly called the root cause. Management's fingerprints were found all over those.

But, had the SOP been followed and the equipment not failed in service, no one would have ever heard of Bhopal. That is the point of the Critical Safety Factors.

Now to the matter of control: do you think management has any *control* over any of these Critical Safety Factors?

Of course they do. Management buys the tools and equipment; writes and approves the methods and procedures; establishes the plan and schedule; develops and delivers the training; decides who's qualified to do the work; determines the crew size and assigns the work to individual crew members. At some level, management has control over every single Critical Safety Factor. Control is absolute. Unfortunately, control does not guarantee that every follower does what they are supposed to do. Follower behavior is the subject of management's collective influence.

The question here involves Organization Power: could any member of middle management—in particular a front-line supervisor—have *control* over any of these Critical Safety Factors? If the answer is yes—even for one factor—that would mean a front-line supervisor can guarantee no one gets hurt by that factor.

That would be power!

STOPPING THE JOB

In most organizations, a front-line supervisor isn't usually the person who writes the policies and procedures, defines the methods used to perform certain work, having signature authority to purchase new equipment and better tools, or can authorize repair of critical equipment. A supervisor may be handed a crew with undertrained or even incompetent team members or forced to operate with an understaffed crew. That is reality, and that reality makes the supervisor appear to have no control.

However, that perception is not reality. In managing safety, the lack of authority to *make* things happen does not mean a supervisor lacks control, the ability to *determine* the outcome for safety. That is because every supervisor has the power to stop the job!

In situations in which the supervisor believes any of those ten Critical Safety Factors are not adequate to perform the work safely, the next step would be to just say no: "We aren't going to start work on this job until and unless we have…" It's known as Stop Work Authority. It's a power that every supervisor has; just about every supervisor has been told to exercise that power when the job's not safe.

Stopping the job—or not starting it—is control: do that, and it guarantees the outcome, at least as it relates to safety. No, the work does not get done, but nobody gets hurt. Not getting hurt is always the most important objective. It's the Case for Safety.

With a complete understanding of organization power, let's return to the opening scenario, where the customer was pressing a contractor to undertake work the contractor did not consider safe. It serves as an illustration of Upward Leading: the contractor is attempting to influence a customer to do the right thing. The customer isn't buying. First, there are several Critical Safety Factors in play:

- the absence of needed tools: a scaffold and a mobile crane
- the absence of needed skills: a qualified electrician
- the pressure of schedule: the customer wanted the job done by Friday
- the absence of a workable plan: given the customer's demand, the original plan could not be followed

As to who's holding the power, it should be obvious. The contractor is attempting to *influence* the customer, but the customer only has the *power of influence*. The contractor has the *power of control*: the tools and people doing the work belong to the contractor, not the customer. The contractor can—and should—simply say no: "We will not agree to do this job in a way that we do not believe is safe."

That would leave the customer very unhappy, and come time to pay the bill, there might even be a dispute. But everyone would go home alive and well at the end of the day.

As one leader put it, "Better to lose a customer than lose a life."

THE POWER OF INFORMATION

The function of leaders in the middle of the organization is to link those at the top with those who do the work to create value. Middle managers perform that role first by determining what the leaders at the top want to happen, and then by converting

MIDDLE REALITY

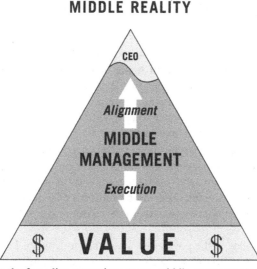

Figure 19.4   From the front-line supervisor on up, middle management always plays to two different audiences: its leaders and its followers.

that desire into action that produces results. In the twenty-first-century language of business, the upward-oriented function is known as *alignment* and the downward-oriented function is *execution*. Getting aligned and managing execution sums up the role and mission of every middle manager (see Figure 19.4).

In practice that means middle managers always play to two audiences: their leaders and their followers. In theory, since these audiences are working for the same organization, they should have the same goals, values, and perception of reality. In practice, they seldom do.

That creates the great plight of middle managers: they're not just in the middle. They serve two entirely different constituencies, which often have little in common, and sometimes have competing goals and values. A middle manager feels the top and the bottom hold the power, and they are powerless. In trying to satisfy the interests of both they wind up pleasing neither, appearing weak and ineffective, and often are caught in the crossfire.

The challenge was explained in The Middle Dilemma. Barry Oshry described "the middle position as being a confusing one. Middles tend not to have clear and firm positions on issues." But Oshry recognized the power of the middles comes from being in the middle: "they have the opportunity to see, interact, and understand" what's really going on in the system. "It is this contact with, information about, and familiarity with various pieces of the system which provides Middles with their power."

Having that knowledge about what's really going on is a huge form of power. Nobody in business knows more about the condition of those Critical Safety Factors and the performance of those working at the value line than do the front-line leaders.

ORGANIZATION POWER IN PRACTICE

Now you understand the real truth about power in the organization as it relates to managing safety performance. Absolute power is found in control, the ability to determine the outcome. By virtue of being able to stop any job where a Critical Safety Factor is not adequate, a front-line supervisor has control, the best kind of power. Moving up the hierarchy (in the conventional wisdom) managers and executives do not have that kind of control; their power comes from their ability to influence the person in control, the supervisor.

On the other hand, when it comes to the choice of behavior by those doing the work, a front-line supervisor has no control, only influence. But of all the members of management, the influence of a front-line leader is direct, and therefore the most powerful, and their credibility is likely to be the highest. Moving up the hierarchy and away from those doing the work, leaders have less influence—not more.

Knowledge adds to the power any leader has to Upward Lead. The boss operates with less information about reality: the competency of those performing the work, the adequacy of tools, methods, procedures, and equipment, and the choice of behavior of those performing the work. A front-line leader has the best view of that reality. Instead of spending a lot of money on a safety survey, executives would be better off asking their front-line leaders, "What's really going on out there?" Of course, that approach only works if the front-line leaders actually tell them the truth. It can be tempting not to.

Bottom line, despite appearances, a front-line supervisor has enormous power over safety. That power is exercised by control—over such things as tools, methods, procedures, equipment, competency, and qualifications—and by direct influence on the choice of behavior by those doing the work. When it comes to managing safety performance, a front-line supervisor is *the most powerful* member of management in the enterprise.

But it seldom feels that way. And that leads many front-line leaders to act as if they have no power. When leaders act as if they have no power, it amounts to the misuse of power. Decisions still get made, but they're the wrong decisions; problems go unresolved, and followers' behavior isn't properly influenced.

It can be a recipe for disaster.

LEADING FROM THE MIDDLE

There are occasions in which the exercise of Organization Power—control and influence—requires Upward Leading where the target of influence is the boss. Some managers welcome this kind of escalation, but there are the tough cases—the "don't bring me problems" types—that call for a leader to possess equal parts courage and diplomacy.

8:15 a.m. Friday Morning, TDB Department

It had been a long night for Ed. As the technical advisor for the TDB Department, he had to lead the troubleshooting effort on the unit's A 21 compressor. By the start of the day shift on Friday morning, that problem was solved. Now the challenge was to bring the rest of the equipment online, something that was never easy in a 50-year-old unit.

"Technical advisor" was the title that Plant Manager Joe Black had bestowed on the few former front-line supervisors who decided to stay when their jobs were "reorganized" a couple of years back. "A fancy name for jack of all trades" is how Ed saw it. But with two kids still in college, he had decided to take the assignment. Besides, after 15 years of being responsible for anything that went wrong, he found that serving as "advisor" had its appeal.

Being the technical advisor in the TDB unit only sounded easy. TDB was one of those no-growth businesses, meaning the unit was under-engineered, undermanned, and under-maintained. "The Salvation Army," as it was known around Acme. But it was still very profitable, so the pressure for production was always significant, leading to this morning's challenge.

Even though part of the instrument system was not in service, Ed's boss authorized the start-up. That was standard practice: back in the days when Ed and his peers were operators, they had gotten plenty of experience doing that. But the current operating team "empowered to run the process"—another one of Joe Black's great ideas—didn't have his experience. Not that it stopped them from being "qualified" to start up the process.

"Qualified. Just another piece of paper proving nothing," Ed thought. If management had known the real story about these guys, they wouldn't sleep at night. "What do the people up front know about what really goes on around here?"

Sitting at his desk, Ed was deliberating his next decision: whether start-up should continue in his absence, or be postponed until Saturday. Having already worked a day shift on Thursday, and called back at midnight Friday, he knew that working all day Friday and well into the evening wasn't a possibility. As technical advisor he could delay the start-up, go home, get a good night's sleep and personally take the lead in the start-up effort on Saturday.

That was the common-sense decision.

But doing that would delay getting back on line for at least another day. The customer was screaming for the product. Complicating things was the reorganization. The operating team concept required full rotation by all operators among all four of the operating positions. That served only to aggravate the experience gap: operators had to work every post to qualify for the Acme Star Point Bonus Program. Another of Black's brilliant ideas. The reorganization was his baby, and he had all kinds of performance metrics to show how well it was working.

On paper, the A Shift Operating Team was qualified to restart the unit, whether Ed was there or not. Their boss—the department manager—had signed off on that. "A wise political move," Ed had thought at the time. But with only one senior operator on the team, and three relatively inexperienced operators, it wasn't a good situation. Start-ups can pose the kind of problems that operators don't normally deal with as part of their training.

Over a cup of coffee, Ed thought about his options:

- As the technical advisor, he could announce the postponement of the start-up and lead the effort on Saturday.
- He could leave well enough alone and just go home. Everyone up the chain of command had signed off that the team was qualified to do the start-up without him.
- He could escalate the problem with his department manager.

> Ed knew exactly what would happen if he chose the third option. It wouldn't be ten minutes before he was sitting in Joe Black's office. Come crunch time, his boss was never willing to make a tough decision, and this one was certainly that. His department manager was skilled at "facilitating dialogue"—a polite phrase for letting someone else take the heat. After a long night's work the last thing Ed wanted was to be sitting in Joseph T. Black's office, facing the grilling that came with raising a problem. Black was notorious for pointing to the sign on his desk: "Don't bring me problems. Bring me solutions."
>
> Reluctantly, Ed got up and headed down the hall to his boss's office. "This isn't going to happen on my watch."

Fortunately for this operation, its employees, shareholders, and customers, and ultimately for the people running the company, Ed did the right thing: he said no. That wasn't the easy choice, and not the decision every peer would have made. The investigation reports of headline-making events stand as evidence that many leaders find it easier to go along, taking their chances they won't live to regret the decision.

In the heat of battle, what middle managers often fail to appreciate is that allowing something like a start-up or launch to proceed when there are serious misgivings and fingers crossed constitutes both a decision and the exercise of control. Saying no means not giving the OK—and therefore determining the outcome. That's not how most supervisors see it. Thinking they don't have any power, they don't Upward Lead, and go along with what they see as someone else's decision. When that happens, a leader is actually mis-leading their followers, putting them in harm's way.

Bearing the brunt of the criticism likely to be forthcoming from their leaders can cause the best middle managers to duck making the tough calls—even when the matter involves safety. In one of those real-life situations, late one January evening in the 1980s, an executive for a contractor pressured his followers to go along with what the customer wanted, telling them the equivalent of "You've got to stop thinking like a technical advisor, and start thinking like a manager." So they did as instructed.

Those words were said, no doubt by a well-intentioned leader, in a pressure-packed situation not unlike either of our scenarios. If someone who properly understood their Organization Power had simply said, "No, not on my watch" the ill-fated decision to approve the launch of the *Challenger* would not have been made.

That's hindsight. What's needed in real time is an understanding of Organization Power—and courage.

IT'S NEVER EASY

Don't think for a moment Upward Leading will be easy. Lead from the middle, it's as if you're wearing a bull's-eye. Two, actually, as leaders in the middle run the risk of being shot at from above and below. Leaders—particularly good leaders—make very convenient targets. First, because they stand for something. In the case of safety, that something is the most important objective a business has: seeing to it that everyone goes home safe. Followers look for that quality in leaders, and admire it when they find it—some of the time. That's because a leader's belief in something—talking a good game—is not nearly enough. Leadership demands action; but when leaders take action, they're bound to ruffle a few feathers. That explains the second reason why good leaders make such convenient targets.

In the long run, doing things the safe way is always good business. In the short run that's not always the case. Doing things the safe way slows things down, takes more effort, and might even cost more money. Moreover, insisting that things be done the safe way doesn't always sit well with followers: it's an inconvenience, it's uncomfortable, it's not the way they want to do it. That's why leaders get resistance, even for something as important as safety. That resistance may come from below—from followers, or from above—the leader's leaders.

Leaders we admire didn't have it easy: Lincoln, Churchill, King, Mandela, John Paul. At some point, many suffered for what they believed in: they may have been roasted by public opinion or threatened with being fired. At the worst, they may have been shot at or put into prison. Did they know that was what they'd face going in? Maybe yes, maybe no. Did that stop them from standing up and doing what they thought right? Absolutely not.

There is one final thing the leaders we most admire have in common: they succeed. The world is a different place because they believed in something important and acted on that belief. That's why we call them leaders—and admire them.

Viewed in that light, leading from the middle—particularly in the matter of safety—isn't all that tough. Nobody's ever threatened a supervisor with prison for stopping a job because it wasn't safe. Leading people to work safely just seems tough because every leader wants to be liked by their followers, and appreciated by the boss.

Here's one last example of a leader who didn't let a little thing like his popularity with followers stand in the way doing the right thing for their safety. The time: 1969. The place: the Chu Lai region, Vietnam. The leader: Lieutenant Colonel Norman Schwarzkopf.

In his autobiography, *It Doesn't Take a Hero*, Schwarzkopf described the kind of resistance he faced when managing the safety of his troops. In those days the meetings designed to allow the troops to air their grievances to their commanders were known as *rap sessions*. The leader would sit in a chair in front of the troops and they'd fire away with questions and complaints under the theory that it was good for morale.

Those rap sessions were never fun...The men would ask, "Why do we have to go out in these minefields? Did you volunteer us for this mission? And, "Why do I have to wear my helmet and flak jacket all the time? They hated the minefields. They hated the heat. They hated their helmets and flak jackets. Most of the time they hated me. But I never made the mistake of confusing their comfort with their welfare. I'd say, "Look, guys, I ain't here to win a popularity contest. My primary concern is keeping you alive. If, on the day you leave for the United States, your last thought of me is "I hate that SOB," that's fine. Actually, I'd be *happy* if that happens. Because an alternative for you is to go home in a metal casket, and then you won't be thinking anything at all. That's why I make you put on your helmet and flak jackets.

Leading is never easy. But leading people to go home safe is always a leader's most important duty.

CHAPTER 20

MISTAKES MANAGERS MAKE

Organizations don't have memories. People have memories.

—*Trevor Kletz*

In life there are two ways to learn: the easy way and the hard way. The hard way is to learn from your mistakes. We all experience life firsthand, and everyone makes mistakes. Supervisors, managers and executives also spend a lot of time dealing with the mistakes their followers make. It comes with the job. Like their followers, leaders are human and make their share of mistakes, too. The mistakes leaders make look different—because they are different: leaders aren't the ones with their "hands on the tools." Their work is to plan, lead, organize, and correct. If that sounds like continuous improvement, it is not a coincidence.

The problem with learning from the mistakes you make as the leader is their cost: yes, you will learn—and learn a lot more from failure than from success—but making a big mistake can ruin the lives of your followers and wreck your career. The easy way is to learn from the mistakes of others. The tuition is free, but there is a prerequisite: you have to find out all about them.

All too often mistakes made by managers are quietly buried ("The COO is leaving the company to spend more time with family"), referred to as the unintended consequence of change ("In our focus on global competitiveness, we lost focus on basic execution"), or spun into some kind of mixed success ("The failure was a by-product of our can-do culture"). Sometimes a failure is referred to as a footnote in the annual report or in a brief admission in an investigation that is otherwise devoted to the bloody details of a tragedy. Management failure is commonly explained away, by using popular terms such as "culture" or "organization silos" or "lack of oversight" or "lessons learned" which explains nothing about what really did go wrong, and who did something wrong. That's probably not a coincidence.

To learn from your peers, you must first be willing to invest your time in finding and understanding the mistakes they made, and you must admit you're capable of making the same mistake yourself.

Unfortunately, leaders are often unwilling to do either. They're too busy to take time to study the failures of others. They think, "That'll never happen to me" as if to suggest those responsible for making mistakes were incompetent. "That can never happen here." But those mistakes did happen, to other leaders in other operations, and they probably thought the same thing.

The smarter strategy is to be a student of other leader's failures, and to presume, "There but for the grace of God go I." The safety historian Trevor Kletz noted the same failures were often repeated on a thirty-year cycle. Why three decades? Simply because those who were there when it happened and survived learned *their* lesson. Eventually they moved on, leaving their successors to learn again—the hard way. Kletz wrote, "Organizations don't have memories. People have memories."

Unfortunately, there's a nearly unlimited supply of mistakes to draw upon. Our interest is on mistakes that impact safety performance, and in particular those involving management processes. A great deal of the content of this book came from leaders who learned their lessons the hard way. By comparison, you need to only read—and study. In this chapter, we'll focus on mistakes that don't fit within the specific topics found in the previous chapters, but have lessons well worth learning.

Learn these lessons well, and you will have a huge advantage. Ultimately, your followers will be the biggest beneficiaries of your knowledge (Figure 20.1).

Finally, as you reflect on these mistakes, you would do well to form your own personal philosophy on mistakes, no matter who makes them. How do you think about them? How do you treat them? Your mistakes may seem different, but in principle, there is no difference between the mistakes made by followers and their leaders.

MISTAKES MANAGERS MAKE
MANAGING SAFETY PERFORMANCE

✓ Failing to prepare

✓ Driving out all fear

✓ Focusing on the short run

✓ Trying to buy a game

✓ Confusing perception with reality

✓ Communicating without a common vocabulary

Figure 20.1   Top ten mistakes managers make managing safety performance.

MISTAKE NUMBER 1: FAILING TO PREPARE

When you fail to prepare, you're preparing to fail.

—John Wooden

Very few people begin their industrial careers in a supervisory position, responsible for the work of others. For most, the first job in business is as an individual contributor, be it apprentice, operator, draftsman, engineer, or accountant, where the one they are responsible for is themselves. Not that learning a new job and beginning a career doesn't present its challenges, but managing people requires an expanded set of skills that very few are equipped with when they are first hired.

As their careers progress, many find out they like what they are doing; some become really good at what they're doing. For them, doing their job doesn't seem all that difficult. It doesn't take long before they are recognized for their skills, and for a few, their leadership potential.

Then one day someone offers an opportunity to manage others. Whether that was what they wanted to do all along or something that never crossed their mind, they're flattered. They accept the promotion and start a new career in supervision.

Sound familiar?

Of course it does: that's how you became a leader.

MANAGING OTHERS: THE GREAT CHALLENGE

Of all the new assignments anyone encounters over the course of a career, no job change is bigger than the one that takes someone from managing themselves to managing others. In industry, that assignment comes with a reminder, "As you know, you are responsible for the safety of those reporting to you."

Sure. It goes without saying. Everyone knows being accountable for the safety of others is part of the job. But do they really understand what it means? Do they appreciate they are now responsible for how other people behave—whether they are standing next to them or not? That this responsibility can weigh so heavily on them as they sit with the family of an injured follower, not knowing if they will survive? That this responsibility would require them to deal with people they work with in ways that wouldn't always make them happy to see you?

Rarely are these matters discussed as part of the offer of promotion to supervision. Were they, a lot of candidates might turn down the job. For good reason: managing safety performance is the toughest part of being in a leadership role in operations.

It is also the one for which people have the least preparation.

Picture this:

You're looking for a good athlete from within your workforce for a very special business opportunity. You would like to find someone with great hand-eye coordination and a track record of success in competitive sports. Fortunately, there are plenty of candidates to choose from. You're given a candidate list: former high school quarterbacks, guards on the basketball team, volleyball players, baseball pitchers, and even a tennis player.

You start interviewing the candidates, searching for the one with the right potential for this special assignment. You concentrate on those who have stayed in shape and kept up their skills. Eventually you happen upon the perfect candidate: a good follower who's been working for you for 10 years and who is a former baseball pitcher who now competes in triathlons.

You offer the job, and they accept. Their new assignment, by the way, is to play golf with Tiger Woods next Monday morning in front of a gallery composed of your company's president and hundreds of senior leaders. No matter that the candidate you've selected has never held a golf club in his hands in his life.

Sounds crazy?

Sure it is. But in a sense that's exactly what happens when someone is first promoted into supervision. They're handed responsibility for the most important job there is—managing the safety of others—even though they have no management experience. Someone with the potential to learn a new set of skills—leadership and management skills—is put into a situation in which he or she is expected to be able to immediately perform those skills and perform them so successfully that no follower goes home hurt.

"THEY'LL DO JUST FINE"

You wouldn't send out a machinist to troubleshoot a problem with electrical switch-gear. At least not without some assurance about their electrical qualifications and training. But you will promote that same machinist to supervisor and expect they will be able to manage the safety performance and behavior of their crew. "They'll do just fine" is what everyone says, probably because that's exactly what happened to them, and they managed to survive the experience.

Big mistake. Failing to prepare is preparing to fail.

When it comes to the most important role a leader in industry plays—sending followers home safe at the end of the day—rarely is there adequate preparation given to the newly promoted leader. That was how their manager started off their career; they turn right around and repay the favor by doing exactly the same thing to their followers: selecting people with great potential as leaders, entrusting them with the safekeeping of their followers, yet failing to give the kind of support and training needed to help them function effectively.

It is amazing that there aren't more failures.

If the new leader is fortunate, they'll have an experienced crew with a few peer leaders who will take up the slack and even coach up their newly promoted leader. But what if the crew is inexperienced? What if the manager is too busy to provide the coaching and development the new supervisor needs?

Learn this lesson from those who have made the mistake of failing to adequately prepare new leaders to successfully manage safety performance. It's one of the most serious errors a manager can make.

MISTAKE NUMBER 2: DRIVING OUT ALL FEAR

Early and provident fear is the mother of safety.

—Edmund Burke

What Peter Drucker did for how we manage our business, Dr. W. Edwards Deming did for how well we make our products. This towering genius brought his "fourteen points" to organizations all over the globe, in the process profoundly changing both manufacturing processes and the quality of the goods and services that are the output of those processes. One of those fourteen points was to "drive out fear."

Dr. Deming came to public prominence later in his life. Born at the start of the twentieth century, he was educated in mathematics and physics. He did an internship at Bell Telephone's Hawthorne Works. He graduated at the height of the Great Depression; in spite of the scarcity of jobs, he managed to find work in the United States Department of Agriculture. In the late 1930s, he joined the Census Bureau, where his training in statistical sampling methods proved useful in the 1940 Census.

Then his career started getting interesting. With a world war going on, Deming brought his knowledge of statistical sampling methods onto the factory floor, helping to dramatically improve the product quality of US war materiel. In the 1950s, he was invited to Japan to help rebuild its manufacturing sector. What worked in munitions factories worked just as well in consumer products plants, as Deming proved in Japan.

Fast forward to the 1980s, when industrial companies began to fully appreciate what the Japanese had accomplished in the quality of everything from automobiles to consumer electronics. They achieved that dramatic improvement by following the advice of a statistician who got his start counting cows and people. The Japanese honored Dr. Deming by naming their national quality award in his honor.

A GIANT LEADING THE WAY

When industrial leaders in the States finally woke up to the "quality revolution" that had been taking place in the manufacturing sector, there to help was the towering figure of W. Edwards Deming. He was in his eighties, but still a commanding figure, both intellectually and physically; the good doctor stood every bit of six feet, eight inches.

In his years of working with industrial clients Deming built what many in manufacturing management would learn as his 14 Points of Quality. These were the principles and practices Deming believed absolutely essential for managers to follow to achieve the highest standards of product and service quality. It was a great stuff for managers to pay attention to, and learn from.

In the middle of his list of points was the instruction "Drive out fear." Deming believed that in the campaign to improve quality fear of getting into trouble for making defective products and reporting quality problems was a major roadblock to progress. There was plenty of vivid experience to justifiably lead him to that conclusion.

Management cannot fix what it doesn't know about. Deming wisely concluded the fear of reprisal from management kept many employees from reporting product quality problems. The staff in the warehouse would rather ship a defective product and let the customer figure out there was a problem; surely management wouldn't fire the customer. Of course, having the customer finding the defect was never good for sales.

If Deming's point was right for quality improvement, why wouldn't it be just as good for improving safety? More than a few leaders thought it would, and applied it to safety.

It sounded like a great idea. People would report all their near misses and tell the absolute truth in every investigation. All that was necessary was for leaders to tell them nothing bad would happen to them. Talk about driving out fear!

But a vital step was missing, one that Deming took as he built up his points: thinking critically about the implication of idea. Deming built his list of points over a career that spanned more than 60 years; there was a lot of trial and error, and hypothesis testing. When it came to managing safety performance, leaders from the *One Minute Manager* school figured, "That sounds like a good idea. Let's give it a try and see how it works."

THE PIPER MUST BE PAID

It failed to work nearly as well as they thought it would. In the first place, employees didn't trust their leaders not to take action when they told the truth. The few who did were sorely disappointed in the cases where leaders did not—or could not—keep their word. Sometimes what they told their leaders just had to be dealt with…including their own choice of bad behavior.

If a leader had thought critically about the idea of driving out fear as it relates to safety, their thinking might have looked like this:

> Drive out fear. Fear of what? The answer is *consequences*. It's the consequences that employees fear. When it comes to making a quality product, we wish they wouldn't fear such consequences. If we have a product quality problem, we are better off knowing about it than having our customers find it for us.

> Are the consequences of making and shipping a defective product the same as a serious injury to our employees who make the product?

> The answer, of course, is no (unless the product defect causes injury to someone else). Our forklift operator breaking his leg is far worse than our shipping a batch of paint that doesn't match the color specification. The paint can always be returned. On the other hand, if the forklift operator had a near miss—nearly running over someone—would we be better off knowing about it? Sure we would.

What would prevent the operator from reporting that near miss?

Fear of getting in trouble, which might happen if the operator admitted to driving way too fast and not paying attention when almost hitting a co-worker at the warehouse.

So, in practice our forklift operator makes the calculation of consequences: better to be safe from management and not report the near miss than run the risk of getting in trouble. Getting in trouble is the consequence that people fear. It keeps them from telling management what's really going on out there.

But is the fear of getting in trouble with the boss really the greatest consequence our forklift operator should fear?

Of course not. It's hitting and hurting another good follower. Or getting hurt themselves.

Do we really want to drive out that fear?

The failure to apply that kind of critical thinking was a big mistake. It's hardly unprecedented: the same thing happened a decade later, when safety was defined as a "value—not a priority."

THE REALLY PAINFUL CONSEQUENCE

Give "fear" more than a moment of thought, a far greater fear emerges: the fear of doing serious harm to a co-worker, like that person almost hit by operator of the forklift. How would anyone feel having to go through the rest of their life knowing they had been responsible for the permanent injury—or death—of someone they worked with?

Fearing a consequence like that is a good thing. This is a fear that every leader should want to increase, not drive out. That fear is what motivates people to work safely.

Yes, there are times when a leader is forced to deal with a follower for whom the fear of discipline is their principal motivator. "I'll never get hurt doing that." If that gets them to follow the safety rules and pay attention to what they're doing, it's not all bad. What motivates many to obey the speed limits is the thought that there just might be a cop around that next curve in the road, and they don't want to get a ticket. But if they were sure the speed limit would never be enforced, how fast would they drive?

Trying to drive out all fear is big mistake. Instead of following Dr. Deming on this matter, it's better to follow the advice of Edmund Burke, who advised, "Early and provident fear is the mother of safety."

MISTAKE NUMBER 3: FOCUSING ON THE SHORT RUN

The difference between luck and skill is seldom apparent at first glance.

—*Investment advice from Peter Bernstein*

The management team has gathered around the conference table in an emergency meeting. The urgent topic: what to do to stanch the rising tide of injuries and events?

It's a scene familiar to any line manager who has at it for very long. Safety performance never follows a straight line. Like competitive athletics, safety performance is made up of streaks and slumps. Even operations with the best safety performance hit the occasional downdraft, leading to meetings like the one just described.

These crisis meetings are guaranteed to produce a flurry of activities, all designed to have an immediate and substantial impact on safety performance. They "round up the usual suspects." You know what that looks like: send a letter urging everyone to pay more attention to what they are doing; show up at safety meetings with the same message; call a time-out for safety; solve the specific problem that was the factor in a recent accident, like issuing work gloves with red paint on the first finger or handing out a stop work coin.

Then they sit back and hope it works.

INSTANT IMPROVEMENT

Fortunately, safety performance usually gets better. When it does, it confirms what they knew all along: when the managers get involved in the details of managing safety performance, they do it better than anyone else around! Are they good, or what?

Once they have finished fixing "that safety problem," they get back to working on all the other business problems they face every day. If he were there to watch all this, W. Edwards Deming would be rolling on the floor, laughing.

As Deming understood so well, what the leaders are dealing with is random variation. The esteemed statistical expert would explain the phenomenon: everything in life has variability. When the numbers go one way, they will eventually go the other way. To the statisticians, the term is regression to the mean. It's the mean—the long-term trend line—that matters to the process of leadership, not short-term uncontrollable fluctuations in performance.

Time after time, the phenomenon fools leaders. When they jump in and act, they see results: performance improves. They think they're having an impact on performance. Surely it must be the direct result of their good work…right?

That's what it *looks* like, but the truth is they'd probably have seen the same result if they had all just gone on vacation.

Assuming a short-term safety improvement is the result of whatever instant effort they made is a common mistake managers make in managing safety performance. Worse, it's only half the story.

THE LONG TERM

While managers typically overestimate the short-term impact of our direct involvement, they underappreciate the long-term impact of their performance as managers. The trend line that safety performance regresses to is essentially the measure of their competency and value added as line managers. All along that's how they should have been thinking about safety performance.

Big mistake.

Safety performance varies significantly from one industry group to another. Within each industry group, though, the means and methods to perform the work are roughly equal, the exposure to job safety hazards is comparable, and operations generally draw upon the same pool of human resources.

That being the case, what separates the performance of the best and the worst within an industry group? The answer lies with the collective performance of management, those who lead and manage safety performance. As Drucker put it, "Companies don't compete. Managers compete."

Of course, that is not how most of the managers look at the situation. Those other leaders in the peer group of companies—the ones getting the best results—must have something going for them that we lack. They have a better culture, a better workforce, followers who know to not report injuries, better tools and equipment—or a different program.

If it can't be explained any other way, they must be luckier.

You can hear Dr. Deming laughing.

Industry group comparisons put everyone on an equal footing: in the long term, hazards are cancelled and so is luck. Managers are left to face the fact that collectively they're getting exactly the results they deserve. Over the long term, the numbers prove that true.

The safety performance metrics for an organization reflect the collective performance of the managers who manage it. A comparison of those performance numbers to what the best in the business are achieving is a good measure of just how good the management really is. Better yet, compare those performance metrics to what the best in every business in the world is achieving. That's what world-class performance is. If your collective performance doesn't put your operation at the top of the hill, you can't claim "in our company, we have a great management team."

In managing safety performance, confusing short-term luck with long-haul results is one of the biggest mistakes managers make. Make that mistake, you'll wind up working on the wrong problem.

MISTAKE NUMBER 4: TRYING TO BUY A GAME

> This club is guaranteed to improve your score by 20%.
>
> —From a golf equipment informational

Sooner or later anyone who has ever golfed has succumbed to the temptation: buy the latest club to hit the market...the one guaranteed to knock strokes off next Saturday's round. And every once in a while, the latest technology works like magic. At least for a few rounds...and then performance reverts to form.

Most of the time nothing really changes. Eventually the new club winds up in the back corner of the workshop, where it has plenty of good company with all the other "breakthrough" clubs bought to help play better. After all, lowering the score is the goal of every golfer—just as lowering the injury rate is the goal of every manager.

On a gorgeous autumn day a few years back, a Hall of Fame golf teacher named Bob Toski put on a clinic for 60 of leaders in the maintenance and construction business. Along the way, he asked for a show of hands: "How many of you bought expensive new drivers and putters this year?" Every hand went up.

Then he asked, "How many of you invested in golf lessons?" One poor golfer timidly raised his hand, perhaps embarrassed to admit he was actually taking lessons. Toski glared at his audience: "There's your problem: you think you can get better by *buying* a game. But it doesn't work that way."

Toski was right about playing better golf—and right about the principles of improving safety performance.

CAN YOU BUY PERFORMANCE?

Leaders—particularly executives—are always on the lookout for a quick and easy way to improve safety performance. There's the carrot-and-stick approach: put in a safety incentive system and simultaneously make an example out of the poor follower who got hurt yesterday. Audit as the means of improvement: inspectors, auditors, observation programs. Training initiatives on culture and human factors. Employee involvement: committees and peer appraisals. Vision and value statements communicated by videos and on posters. Back belts. Family pictures on ID badges. The list goes on.

Sometimes the method works. But more often they don't work any better than that new golf club. Why is that?

Buying a safety game meant managers can avoid having to change how they manage. They can just keep on swinging the way they always have, but get different results. The new equipment would do the heavy lifting. Or so they thought.

It doesn't work that way for golf—and it doesn't work that way for managing safety performance. If we want better results, processes have to change, and that requires investing in *improvement*. For golf that means lessons from the pro and hard time on the practice tee. You cannot send somebody out there to practice for you; nor can you buy a lower score with your MasterCard.

When improving operating performance, leaders understand that perfectly well, and engage in process improvement using Lean Manufacturing and Six Sigma methods. It's hard work, led by people who know the science of improvement.

When it comes to improving safety performance, the process can't be different. Getting people working safely is all about *execution*. Improving the way people in the organization execute their work every day requires leadership, and better leadership than has been employed in the past. You cannot expect better results with the same swing in either golf or management.

The route to better leadership is the same as in golf: taking lessons from the pro and spending time on the practice tee. That's investing in improvement, instead of trying to buy a game. Yes, it takes a greater initial investment of the time and energy of the leaders, but over the long haul, it's a great investment.

In managing safety performance, there are no shortcuts to excellence. Don't make the mistake of trying to buy a game.

MISTAKE NUMBER 5: CONFUSING PERCEPTION WITH REALITY

> Perception isn't reality. Reality is reality.
>
> —Joe Girardi

At the press conference where he was introduced as the new manager of the New York Yankees, the former Yankee catcher and one time manager of the Tampa Bay Rays, Joe Girardi was asked a question by one of the New York press corps: "Joe, when you were the manager at Tampa, the perception was…"

Girardi interrupted, not letting the reporter finish the question. "Perception isn't reality. Reality is reality."

It was one of the most useful statements ever made about managing safety performance.

Safety is real; it is grounded in reality. Hazards are things that can harm people. The goal of managing safety performance is to keep hazards from harming the people working around hazards. To do that, hazards must be recognized and dealt with properly. Either people suffer harm, or they do not.

None of that has anything to do with perception. But it doesn't stop leaders and their followers from regularly talking about perception as an important factor in managing safety performance to the point where it would seem that "perception *is* reality."

Big mistake.

Just because someone does not perceive a hazard does not mean there is no hazard present. Just because a leader perceives that safety performance is good does not mean it is.

Perception is found in the five and a half-inch space between the ears. Hazards are found in the real world. It is the cold hard steel of reality that causes harm. To work safely, those doing the work in the presence of harm must recognize those hazards. The leaders responsible for those doing the work must recognize what those hazards are and how well their followers are doing when working in their presence. None of that involves perception. All of that involves recognizing reality for what it really is.

Perception is not reality. Reality is reality. Do not make the mistake of confusing perception with reality.

MISTAKE NUMBER 6: COMMUNICATING WITHOUT A COMMON VOCABULARY

> What we have is failure to communicate.
>
> —from Cool Hand Luke

Actions may speak louder than words, but leaders use words all the time. Words function as the means of exchanging information among us humans, no matter whether we are leaders or followers. Effective communication succeeds in making that exchange: the information sent conveys its intent to the receiver. If not, the result is miscommunication, with all its subsequent ill effects.

When you were in grade school, you learned the meaning and use of the word vocabulary: a vocabulary is the collection of words used to communicate—and the meaning of each word. A common vocabulary is required to effectively communicate; without one, people don't really know what they're saying or hearing.

Just because the communicator and audience learned to speak the same language does not mean they understand the meaning of the words they use. That is something you understand perfectly; personal experience has no doubt taught you that lesson well.

Still, knowing that is not enough. Leaders fail to appreciate the problems *they* create with *their* vocabulary; problems that become serious when safety is the subject of the communication. As if communicating isn't tough enough, leaders routinely make communicating tougher by using words that any follower could look up in a dictionary (not that they would) *and* creating a different meaning for the word. It's *their* meaning.

Making matters even worse, these leaders can't provide a simple, understandable definition for the words they use, the likes of which could be found in any dictionary.

Instead, they communicate presuming "everybody knows what I mean." Rarely do they, no matter who they are, followers or other leaders.

WHAT'S IN A WORD?

In a different time and for a different business function, the same problem was diagnosed by Dr. Deming. There was a time in business when the word quality had almost as many definitions as there were leaders to define it. Product quality could be a matter of appearance, cost, what the Quality Department would approve for shipment or what the customer would accept. As a first step in the process of improving product quality, Deming insisted that there be a commonly understood operational definition of the word "quality." His recommendation: quality should be defined as the conformance to requirements.

It was a stroke of genius.

If you're looking for examples of the same problem as it relates to managing safety, you need not look far, as the preceding chapters of *Alive and Well* have laid out numerous commonly used words that have become terms of art in the process of safety leadership, but are rarely defined by the leaders who use them.

Examples, in alphabetical order:

- Accountable
- Attitude

- Behavior
- Complacency
- Control
- Consequences
- Culture
- Hazard
- Influence
- Leader
- Power
- Priority
- Responsible
- Risk
- Value

In the next chapter, "Driving Execution", you'll find another common term, execution, defined, along with two other related terms of leadership art. The creation of a common vocabulary—operational definitions of key safety leadership terms—is an important objective of this book.

There was a time when the term, "Safety is a value, not a priority" made the rounds in the world of industrial safety. Likely it was imported from the United States Army, whose leaders had concluded that their priorities might change with the times, but their core values should not.

Good for the armed forces, for whom this made perfect sense. But without it being given any critical thinking, the term was adopted by industrial leaders and applied to safety. They meant well, and what they probably meant to say is "safety is so important that, no matter what, it will always be at the top of everyone's list." Those are simple words; not so, value and priority. As familiar as they might be, used in this form and context, their meeting becomes unclear.

A value is "something of the utmost importance; something with intrinsic worth." Safety certainly fits that description, but to say "values never change" is simply not true. Over the course of a lifetime, every person's values change, and often change for the better. It's not even correct to say values shouldn't change; there are many cases when they should.

A priority is "that which comes first." Can priorities change? Yes, and they often do. Should the priority for safety ever change? Absolutely not: it must always come first. So, when applied to safety, the statement "Safety is a value, not a priority" is absurd.

A big mistake. The correct way to have stated this philosophy using the words value and priority might have been: "Safety is of so much value to us all that its priority will never change."

But "Safety will always come first. Period." says it far better.

COMMUNICATING CLEARLY

A role of the leader is to make sense of complexity, to explain and educate, and to create a common understanding among followers. When leaders communicate using words like the ones described above it only serves to confuse and frustrate followers. If you want your followers to take your words seriously, you need to be serious about the words you choose.

Critical thinking is required; a good dictionary helps. It's a big mistake to consider this a matter of semantics, that everyone knows what you're talking about, or to communicate without having a common vocabulary.

CHAPTER 21

DRIVING EXECUTION

If you can't execute well, it doesn't matter what the strategy is.
—*Tom Peters*

Coming as it does at the end of the book, a chapter on execution might seem like an afterthought: with all the important matters properly treated, time to finish off by looking at execution. Nothing could be further from the truth. There's a good case to be made that this chapter should be the first chapter in the book. But the ideas about execution will make so much more sense being built from all that you've learned.

Alive and Well really is a book about execution: safety execution. Execution means everything. We began by identifying the toughest safety challenges leaders face every day. The common element for most of those challenges is the need for better execution; the solutions offered in the succeeding chapters will translate into better safety execution. Nowhere in this book is there to be found any suggestion that the solution to these safety challenges is to create more procedures or launch new programs. Instead, the focus has simply been on things for leaders to do to lead and manage safety better.

The ideas and leadership practices described in this chapter will serve to unite everything you've read in *Alive and Well,* creating a coherent strategy to successfully drive safety execution.

WHAT IS EXECUTION?

As with other words that are part of a leader's vocabulary like *accountability* and *culture*, when *execution* is tossed around in everyday conversation, it's presumed everyone understands what it means. And that everyone understands execution in exactly the same way. But, as it relates to safety, what does execution mean?

It's such a simple question; every leader who uses the word execution should have a clear definition of exactly what they mean by execution. So should every follower, who hears the word used by their leader. Seldom is either the case.

Alive and Well at the End of the Day: The Supervisor's Guide to Managing Safety in Operations, Second Edition. Paul D. Balmert.

If you were to answer that question for yourself, you'll probably have to pause, contemplate the question in the context of safety, consider the possibilities, and then choose your words carefully. Taking the time to do that would be such a wonderful thing: it would make you recognize what you know, what you don't understand, and what you're assuming. But busy leader that you are, you don't have the time to do that. Instead, you proceed on the assumption that you know what you mean, and so do your followers.

Or that it's really not that important for anyone to know for sure.

Were you to take time—you really should—no doubt you'd come up with a good definition, and you'd have a better understanding of what you mean when you use the word execution. The problem would then be that it's your definition; leaving everyone else to operate on their definition of execution.

The better approach would be to adopt a common definition of execution, one that is simple, understandable, and useful. Make execution part of the common vocabulary, there would be one less communication problem to deal with.

There is an operational definition of execution that is exactly that: simple, understandable, and incredibly useful. Even better, it's a definition that can be traced to the same kind of great thinker that was Peter Drucker, who provided the observation that has become the foundation of the Leadership Model. W. Edwards Deming is known throughout the world for his "virtuous cycle" of Plan/Do/Check/Act. Of course, Deming was quick to acknowledge others as the source of the model, and PDCA is nothing more than the Scientific Method put into practice as the means of continuous process improvement.

But, as to execution, Deming observed, "All work is a process." In Deming's model, a process converts inputs into outputs. To improve the quality of the output, Deming focused effort on improving the process and, if needed, the quality of the inputs.

Deming's definition of a process offers the perfect definition of execution: it is the *doing* part of any work process.

The operational definition of *execution* really is that simple. But when you fully understand what those words mean, you will come to appreciate the power locked up by looking at execution simply as the *doing* (see Figure 21.1).

EXECUTION MEANS "DOING"

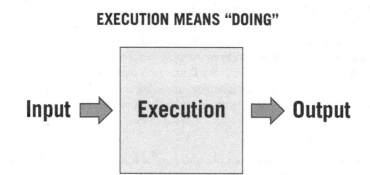

Figure 21.1   Execution ultimately determines the level of safety performance.

For openers, if all work is a process, then execution takes place in some form for all work done. That means everyone plays in the game of execution: those who do the planning, those who do the checking, and those who investigate problems and come up with solutions, doing that work is their execution.

Doing describes execution at the level of the individual. At the level of the firm, those who *do* the work to produce the product or perform the service are the ones creating economic value for the business. Everyone else in the business—sales, marketing, finance, information technology, human resources and management—in some way serve to make their execution happen the way it is supposed to.

The definition of execution implies that things are not only being done, but how well they are being done. Ideally, things are done *right*. Right means that both the letter and the spirit of the requirements are met. Proper execution involves meeting requirements and expectations that may not even be written, and might not even be spoken, such as intervening when someone is taking unacceptable risk, stopping a job that's not considered safe, escalating concerns that aren't being properly addressed.

The scope of execution encompasses an extraordinarily long list of things to be done, from training and qualifying people, inspecting and maintaining equipment, establishing and maintaining procedures, and keeping records. Is there anything on that long list that doesn't have to be done? Done as required? Done well? Now you're beginning to see the magnitude of the challenge: because it applies to everything that is done, the scope of execution is enormous.

Summing up all of what's been said about execution up to this point provides a framework for understanding the answers to the most basic of questions about the nature of execution and the great challenge of driving execution. What is execution? Where does execution take place? Why is execution such a great challenge? Who is responsible for execution?

Consider the answers to these questions, the **Four Absolute Truths About Execution**.

1. Execution is the *doing* part of any process. As such, it's the only part of the process that counts.

What gets people hurt and what keeps people safe is determined by what they do, and how well they do that. Yes, all the other activities—planning, checking, and taking corrective action—are important and help make that result happen. And yes, a poorly devised plan or the failure to define an effective solution to a known problem can lead to an injury. But in the organization sense, those things happen either before or after execution—the doing. Those things are done on paper, by a computer, in an office or during a meeting in a conference room. That's not where most injuries take place. That suggests where to look to find safety execution: where the work is done.

Defined as doing, execution becomes observable. If you want to see what execution looks like, All that is necessary is to look. Execution is found in reality.

When you observe anything being done by humans, it's highly unlikely the doing will be perfect. Nobody's perfect. Rarely is what they do perfect either. Like zero risk, flawless execution isn't attainable: there will always be room for improvement, and that is not necessarily a bad thing.

2. Execution is locked up in the ordinary: the detailed activities undertaken in carrying out what is required.

In the organization sense, execution starts after the strategy has been determined, the decisions made, the plans set. That's often the point at which the decision-makers think the hard work is completed.

In "Turning Great Strategy into Performance" two consultants measured the success organizations had in executing their strategies. The results: not good. Roughly two-thirds of what gets planned and decided actually gets accomplished and converted into results. That's hardly flawless execution. The authors, Michael Mankins and Richard Steele, had this to say about what happens at the doing stage: "The biggest factor of all may be executive inattention. Once a plan is decided upon, there is often surprisingly little follow-through to ensure that it is executed."

Among the reasons for that inattention, execution is found in the details of work far down the chain of command: execution involves what happens out on the job site and shop floor. Improving the level of execution demands paying attention to all the details that happen on every job every day.

If you're an executive, how exciting is it to get into the middle of all those details? Boring might be a better description. Unlike launching a bold new safety initiative, improving the level of execution is not the least bit glamorous. Focus on execution, you won't see performance changing rapidly; when it does begin to improve, it's hardly noticeable. Execution looks exactly like the way things are supposed to be done, right? What's so special about that?

Nothing. And everything.

An operation can get away with doing things that are not "exactly right" for a very long time and never experience any significant consequences. There was no lack of policies, procedures, programs, or standards in the great failures like Three Mile Island, Challenger, Bhopal, and the Macondo Well. Rather, it was the collective effect of shortcomings in execution taking place far beneath the executive level, such as unqualified control room operators, leaking seals, poorly maintained equipment and noncompliance with SOPs that were the at the core of the causes.

3. Execution runs counter to the propensity of natural systems, which move in the direction of disorder.

This follows a fundamental law of physics, inertia: an object will continue in its current speed and direction unless acted upon by an unbalanced external force. Launch a spacecraft, it escapes Earth's gravity, and it will continue on forever, no effort needed. Leaders routinely apply the law of inertia to execution: once a policy or procedure is put in place, they assume it will continue to be faithfully executed unless acted on—by management. That means that once implemented, policies, procedures, programs, and standards require no further investment of the energy of leaders.

If equipment inspections were halted, would equipment be maintained—or would it deteriorate? If leaders stopped enforcing the safety rules, would compliance be maintained—or deteriorate? When management stops paying attention to something, and starts to focus on something new and different, does it go unnoticed? Do followers pay attention to what their leaders pay attention to and follow their lead? Of course they do.

Unless you believe the law of inertia applies to safety, you know that execution requires the continuous engagement and investment of the leaders over time. Both are commodities in short supply in management ranks.

4. Execution is a function of leadership.

All of this suggests that execution is a function of leadership. The word *function* has two meanings. In management terms, function means duty or responsibility; in math function means "dependent upon." Both definitions describe execution perfectly. If execution isn't a duty of management, whose job is it? If leaders don't lead and manage execution, what will be the level of execution their operation achieves?

Managing execution is the job of every leader, from the CEO down. As to the front-line leader, if you are one, you know execution is *your* job. As Drucker said, managers must manage: "taking action to make the desired results come to pass."

As it applies to safety, there is nothing more important for a leader than to make that happen.

THE VALUE OF EXECUTION

Understanding execution as you now do, you would think every organization intent on improving safety performance would see driving execution as the best way forward. That is seldom the case.

What's been offered to managers in the name of safety performance improvement over the last decades? Survey the literature, listen to the experts, benchmark someone else's performance, there seems no limit to innovative solutions: change attitudes, behavioral observation, build a database, buy new software, change the culture, create a management system, promote participation, fix the procedures. There's a certain alphabetical symmetry to these improvement initiatives: attitude, behavior, culture, management system...all on the path to zero harm.

By comparison, in operations Six Sigma and Lean Manufacturing have become core practices used to improve manufacturing execution. If only these improvement processes could solve all those vexing safety execution challenges, companies that build great airplanes, motor vehicles, and buildings would have safety performance to match the quality of their products. They don't. The manufacturing quality improvement process hasn't been able to cause the desired change in safety execution. That suggests something else is missing: leadership.

When it comes to getting better safety results, execution is often the last thing anyone thinks to improve. It deserves to be the first. Why would something as vital as execution be overlooked, not just by the experts, but also by those with the most at stake: senior executives, the ones held accountable for performance? Making a change in execution doesn't require changing policies and procedures or investing time and effort in launching a new program or set of standards.

Driving execution is simply a matter of doing what is supposed to be done— better. Achieving the goal for safety performance—sending everyone home safe at the end of the day—is fundamentally a game of execution.

As to where to focus the attention to achieve better safety performance through execution, the typical front-line supervisor's list of toughest safety challenges reveals the obvious answers:

- Awareness: Getting people to recognize the hazards most likely to harm them.
- Complacency: Stepping up the focus and engagement on the task at hand.
- Compliance: Getting everyone to follow all the rules—all the time.
- Change: Commitment and compliance with new and revised policies.
- Equipment: Maintained and operated properly.
- Experience: New people who don't have enough, and senior people who have too much…and the bad habits to prove it.
- Near misses: Finding out about them, and doing the right thing to solve the problem.
- Production: Getting the work done *and* getting it done safely.
- Risk: Taking the appropriate measures to reduce risk to an acceptable level.
- Stopping the job: People stop working when the conditions are not safe.
- Training: Learning and understanding what's needed to do the job safely.

Driving execution means successfully taking on challenges like these. Succeed and safety performance will surely improve.

THE EXECUTION GAP

By defining execution as doing, in the context Deming's cycle of Plan/Do/Check/Act, the "plan" becomes the standard of performance. While there certainly can be a plan for an individual job, in the broad sense of the organization, for safety the plan is made up of all the policies, procedures, standards, and programs that dictate what people are expected to do.

Defining the plan in that way reveals a huge problem for leaders: what are all those safety requirements? How many of them are there? Where are they found?

They are all written down…somewhere: that's what makes them requirements (although you also have unwritten expectations as well.) Since all involve safety, that means they're all important. They may have originated elsewhere (such as regulatory requirements and laws) but they are your requirements, not your followers'. And you are expected to see to it that your followers comply with all of them. That is the function of the rules.

The reality is that nobody knows the answers to these questions. The number of rules and requirements that apply to one individual follower working in operations might run in 10s…100s; for some, it might be as many as a thousand. But they are the rules; followers are expected to follow all the rules all the time; doing what's expected is the essence of execution.

Unless execution is flawless—it never is—there will be an **Execution Gap** between what's expected and what's being done. The size and nature of the Execution Gap reveals the work to be done to drive execution.

In theory, it would be possible to know exactly what the Execution Gap is: all of the requirements are documented; a wall-to-wall audit conducted by an army of auditors could measure the degree of compliance for every single requirement. You can only imagine what the audit report would look like. In practice no leader would ever undertake such a project. Still, there is a benefit to thinking about the exercise: as a leader, particularly at the front line, you probably have a sense as to the size and the scope of your Execution Gap.

On the bright side, the gap will tell you what the potential for performance improvement locked up in execution is: most likely it is huge. How much better would performance be if every follower simply carefully comported him- or herself according to the plan that is already in place? Followed all the rules, all the time? Paid careful attention to what he or she is doing at every moment? Maintained equipment and facilities up to the standards that have been set? Investigated events and fixed problems appropriately?

The path to zero harm lies with closing the Execution Gap.

PERFORMANCE VISIBILITY

Given there's an Execution Gap for your area of responsibility—remember, the gap applies not just to people, but equally to things such as tools, equipment, raw materials, finished products—there's the question of how much you know about that gap. That's called **Performance Visibility**: the degree to which a leader knows reality for what it really is.

The Execution Gap and Performance Visibility are inseparable. There is always a Performance Gap; a leader will never know everything there is to know about execution and the Performance Gap. How much the leader knows about the gap can range from very little to almost everything. In that way, Performance Visibility can be measured on a scale of 0 to 100%.

As can execution.

As the leader, the more you know about execution and the Execution Gap, the better off you'll be, but attaining a high level of Performance Visibility is never easy to come by. Found in every organization are forces and factors that limit a leader's ability to see reality for what it really is. They are the byproduct of how people in organizations are naturally inclined to behave – yourself included. Every leader at every level needs to understand this aspect of the organization and organization behavior, and in turn, know how to manage the challenge.

If you're the newly hired leader, coming in from the outside, on your first day, as you're being introduced to your new followers, you're Performance Visibility starts out at zero. You know nothing. Knowing that to be so, you're motivated to get out and discover everything you can about reality. You know you can't learn everything you need to know just by reading the reports.

The bigger problem shows up after you've been there so long and have seen so much, you're sure you know exactly what's going on. You assume you have high Performance Visibility. You might be right, but if the phone rings and you're informed of a problem you did not know existed, that's the point when you find out your Performance Visibility is not what you think it is. You can only hope when the phone rings, the problem isn't serious.

An unrecognized factor in most high-profile industrial events was exactly that: low Performance Visibility on the part of upper management, who perceived their safety execution was much better than it actually was. Had they known the truth about the Execution Gap in their operation, surely they would have addressed the problems. Instead, they were shocked.

For example, you may be a front-line supervisor responsible for a crew of repair technicians scattered across a wide swath of customer facilities. As a site leader, you might be responsible for managing two different manufacturing facilities, separated by an hour's drive. A director of operations, you manage a division with facilities on four continents. Reporting to the vice president of production are leaders of dozens of plants employing thousands—and an equal number of contractors. But even for a small facility and a small company, there's a day shift and a night shift. The headquarters staff works across the street. At the start of the shift, the maintenance supervisor hands out assignments that send crew members in many different directions.

Those are the organizational and physical constraints limiting Performance Visibility. Then there are the natural behavioral tendencies of the humans working in organizations. That's what creates familiar situations: not to report near-misses; not to comply with the rules when the boss isn't present; not wanting to make the boss look bad; not wanting to look bad to the boss; to put the best foot forward; to make the numbers look good; to create a "parade route" for the annual site visit by the CEO.

Factors like these—and others—collude to produce a distorted view of reality about execution: low Performance Visibility. Absent evidence to the contrary, leaders assume things must be going great. Low Performance Visibility robs a leader of the opportunities to find out about problems and fix them and to learn from bad experience; it sows the seeds of greater failure, and sets up the leader for a phone call with shockingly bad news.

It's not like leaders don't know that. The problem is the path of least resistance is to go along with that (Figure 21.2).

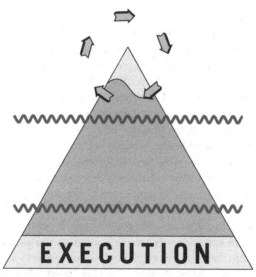

Figure 21.2   In every organization, there are natural barriers that must be overcome to increase the Performance Visibility of safety execution.

DRIVING EXECUTION

Now that you understand the truth about execution, all that remains to be done is to make execution happen the way you want it to. To drive execution, what do you do, and how do you do it?

As a practical matter, the answer depends on your leadership role: the tasks falling to the CEO and the front-line leader are vastly different. No matter what your leadership level and role, an impressive array of techniques is at your disposal. If you want to look at driving execution at the individual follower level, the techniques can be thought of as beginning with training and ending with performance appraisal and compensation, with coaching sandwiched in the middle. If you're looking at driving execution for operation or a business, there's process improvement, design, and human factors engineering. Any and all of these processes can be used to drive execution for a business.

At this point, you have the leadership practices you have learned in twenty-one chapters of *Alive and Well*. While they have been focused specifically on the goal of sending everyone home, alive and well at the end of the day, in some form, every technique and practice can also be applied to improving execution in every business function.

Successfully driving execution will reduce the Execution Gap, but accomplishing that also requires increasing Performance Visibility. You must know reality, and you must be able to see reality for it really is to able to observe and measure improvement in execution.

Well executed, many of the tools and techniques you have found in *Alive and Well* will reduce your Execution Gap or increase your Performance Visibility; some will work on both goals (Figure 21.3).

If you're a front-line supervisor, your job is to manage execution. Here are examples of tools and techniques that you can put to immediate use to do that.

- **Managing by Walking Around**

If execution is what really happens, there's no better way to see execution than to Manage by Walking Around: look closely at what people in the organization are actually doing. Getting your well-trained and sharp eyes on reality will simultaneously increase your Performance Visibility.

As to where to focus your time and attention—the "calibrated" part of MBWA—rules that are the toughest to get full compliance make an easy target. If the statistics and records indicate that driving safety is a particular "execution problem"— for example, if your people are not wearing seat belts, driving excessively fast, or not focusing on the task of driving—what better thing for a leader to do than spend some time on the road with individual followers? Doing that both makes a statement and gives the leader some first-hand data about driving practices.

- **Correcting Non-compliant Behavior**

Found at the core of execution is compliance. In the case of policies and procedures, proper execution requires following all the rules all the time. The SORRY model for correcting behavior offers the best practice to change behavior for the better and do so permanently.

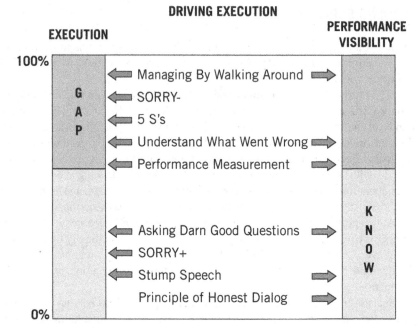

Figure 21.3 Safety leadership tools and practices well executed can increase Performance Visibility and reduce the Execution Gap.

- **Recognizing and Reinforcing Compliant Behavior**

As detailed in Chapter 8, "Behavior, Consequences—and Attitude", the benefits from positive reinforcement of safe and compliant behavior are huge. The SORRY model for reinforcing safe behavior offers the best practice to give positive feedback that will influence future behavior.

- **Understanding What Went Wrong**

Looking into small problems in a way that produces understanding is a simple way to gain a significant amount of Performance Visibility. That's one of the benefits of reporting near misses. But these near-misses need to be reported in order to be able to investigate them. Fixing problems found in investigations decreases the Execution Gap.

- **Your Stump Speech**

The purpose of a Stump Speech is to give your followers your best advice, philosophy, expectations, and values on important subjects. When those subjects involve safety, it's a Safety Stump Speech.

In your inventory of Safety Stump Speeches should be several about execution: What's your Stump Speech on reporting near misses? Being honest in investigations? Following all the rules all the time? Bringing problems to your attention?

- **Making Change Happen**

Communicating a new or revised policy and procedure creates a Moment of High Influence. As you know, there are two simple rules to follow to up the likelihood of successful change: let people know the reason for the change, and then focus on making the change happen.

Making that change *is* execution.

There's a world of difference between broadcasting a policy change by e-mail and giving people a face-to-face explanation—and then listening to what those affected see as the problems standing in the way. That approach to communication might not be the norm, but it's not unheard of. Tyco's CEO Edward Breen did just that, communicating the change in the aftermath of the scandal that put his predecessor in jail. The new CEO made a point of visiting many Tyco facilities all over the world, explaining the company's new ethics code. But it wasn't a one-way communication: the time allowed for questions was at least as long as his presentation. That approach demonstrates the engagement of the leader, and works to generate the same kind of feeling in the followers.

Communicating well requires not just listening, but empathetic listening. Moreover, it means providing real answers to legitimate questions and devoting resources to the real problems created by the change.

- **Darn Good Questions**

Asking questions—Darn Good Questions—can be a very powerful intervention in managing execution. Recall that Darn Good Questions aren't about finding out information like "What time will you be finished with the order?" but have a more fundamental leadership in mind. "What are the problems you're experiencing trying to implement the new safety policy?"

Given that execution involves followers, asking those who actually make it happen about their experience can stimulate thinking and, in the process, provide some eye-opening insight about execution.

- **Managing Accountability**

Found at the core of the 5 Ss to manage accountability are Darn Good Questions: they create understanding and acceptance of what was done wrong; an appreciation of consequences, actual and potential; clarity of duties and expectations; the recognition of potential consequences in the future.

When someone is held accountable, it's likely because execution came up short of the mark. The goal of the process of holding a follower accountable is to change execution for the better.

- **Measuring Safety Performance**

Good metrics can help. Execution is what happens and how well it happens. Getting an accurate picture of that reality should be the goal of performance measurement.

As was explained in Chapter 17, "Measuring Safety Performance," many measures used to *reveal* performance are used as the means to *reward* performance. When a measure like the injury frequency takes on the significance of the scoreboard at the end of the game, the pressure is put on managing the number instead of managing the results. Separate measures are required to provide accurate information as to the state of safety reality. These measures of Performance Visibility can't be allowed to become the basis for any consequences, positive or negative. If they do, it's terribly difficult for followers not to massage the numbers to make them appear better than they are. Doing that may meet their use for performance evaluation but renders the measure meaningless in creating Performance Visibility. That leaves leaders flying blind—but they don't think they are. That's more dangerous than actually flying blind!

One alternative approach is to create measures that allow a leader to look into "the process of execution" as opposed to the output from execution. For example, that might lead to looking into *how* training classes are taught, *what* characterizes jobs that are stopped because they are not seen as safe, *how* equipment is inspected, *how* a Job Safety Analysis or Work Permit is written; *how* an investigation is conducted.

- **The Case for Safety**

If you need any motivation to take on the Execution Gap, recall the Case for Safety: why safety is even more important than any other business objective a leader has. The explanation lies in answers to three questions: (i) "What are the most important things in my life?" (ii) "How would a serious injury to me affect the really important things in my life?" and (iii) "Is there anyone I supervise who has fundamentally different answers to those two questions?"

Taking on execution is the ultimate challenge; those answers will provide the leader with the energy and appetite to pursue safety execution as if someone's life depends on it. It does.

THE PRINCIPLE OF HONEST DIALOG

There can be an alternative route to upping your Performance Visibility: getting a healthy dose of reality from your followers without needing to ask. Nobody knows reality better than they do. But in order for them to "clue you in" two conditions must exist: they must see something in for themselves and they must be convinced no harm will come to them from telling you what's really going on.

So many followers have had a bad experience in "giving up" information, they're highly disinclined to tell you what you need to know, however unpleasant as the information might be. But they might be convinced otherwise if they thought things would get better.

That leads to the final tool in the chapter and in *Alive and Well*. It's called **The Principle of Honest Dialog**, and its definition speaks for itself: when it comes to safety and execution, leaders and followers owe it to each other to be honest.

The cost of doing otherwise, measured in pain, suffering and loss of human life is unacceptably high.

EXPLODING THE MYTHS

Now that you've engaged in detailed examination of execution, it's time to explode **Four Myths About Execution** once and for all time.

1. Deciding is the hard part; executing is the easy part.

There's a presumption by executives that making the decision is the tough part of the work of making change happen. Once the decision is made, it's simply a matter of execution, which falls to others.

Yes it does. For safety, the duty of making change happen commonly falls to the front-line leaders. As was described in Chapter 10, "Making Change Happen," an approved policy is nothing more than a piece of paper with a signature on it. The change does not take place *unless* and *until* those words are converted into behavior. That is *the* change.

That's where the real work of change happens, and where resistance to change is most likely to be encountered.

As the saying goes, "If you think something's easy, try doing it yourself."

2. Execution is someone else's job.

Larry Bossidy, the other coauthor of *Execution* and a former CEO, suggests doing otherwise: "Many people regard execution as detail work that's beneath the dignity of a business leader. That's wrong—it's a leader's most important job." Those details constitute the very nature of execution: the work required to convert a plan into action.

Execution, the *doing* part of that process, is largely ignored by top executives. At least they can say they have plenty of good company: the curriculum taught in business schools seldom includes managing execution; research about execution is largely ignored; books written about functions such as sales, strategy, information technology, and leadership line the shelves, while those on execution are rare.

In "Three Reasons Why Good Strategies Fail: Execution, Execution, Execution" the authors at the Wharton School of Business publication observed that "despite the obvious importance of good planning and execution, relatively few management thinkers have focused on what kinds of processes and leadership are best for turning a strategy into results."

3. Once implemented, programs maintain themselves.

Is it sufficient for a senior leader to communicate the strategy, announce the decision, broadcast the new policy, and then expect compliance—forever? Every new policy, procedure, program, initiative is added to the workload of an organization that is already overburdened. What about all the other policies and procedures, programs and standards put in place in previous years?

Once created, requirements can't be presumed to continue to be executed without any particular investment of the time and attention of the leaders, from the front-line leader up.

4. The human error rate approximates zero.

Do well-intentioned people always do everything exactly the way they're supposed to? Of course not.

When it comes to doing things, how good are we humans? The short answer is, "Not that good." There are any number of useful studies that have calculated the "normal human error rates" for a wide variety of tasks, from entering the right numbers in the computer to correctly troubleshooting the cause of a technical problem in a control room. As you might expect, error rates vary widely, and depend on factors such as the person performing the task (novice or expert), the task itself (easy or complex), and the context in which the task is performed (routine or emergency).

Generally, numbers make intuitive sense: for example, the odds of safely landing a commercial airliner are far and away better than the odds that the passengers' luggage will all show up at the baggage station. Sometimes the numbers are paradoxical: when the pressure is on and time is of the essence, error rates can be very high. Take the pressure off, and the error rate for performing routine tasks can still be very high. It's called complacency.

Expected human error rates have been factored into how systems are designed. For example, the two pilots on a commerical flight crew are required to cross-check and verify each other's work. That logic is applied to activities with potentially life-threatening consequences: a permit for a confined space is created, reviewed and countersigned; work permits are created independently by one person and issued to the second person. In the case of error reduction, two heads are better than one.

On the other hand, risk management protocols and risk estimates are usually made presuming that "everyone will follow all the rules, all the time." It can even be the leaders who fail to do that, by signing waivers and allowing exceptions. That the rules are not always followed creates an additional level of risk—the probability of failure—that can be thought of as Execution Risk.

BACK TO THE BASICS

No matter the level, execution requires a leader's staying power: continuing— relentless would be a better word—focus on those details, day after day, week after week, month after month. As soon as the leaders take their eye off that prize, the rest of the outfit will soon follow. This relentless focus is hardly the norm: even the best leaders gravitate to the squeaky wheel—today's problem or crisis. When safety performance appears to be under control the temptation can be too great to resist. One very fine plant manager took to wearing a rubber band around his wrist to remind him not to fall victim to that temptation. Stretching it, he would note, was required for the band to have energy, and the source of that energy was his leadership.

Finally, for good reason, *Alive and Well* is subtitled *A Supervisor's Guide to Managing Safety Performance*: no leader plays a more important role in managing safety performance and driving execution than does the front-line leader.

If you're a front-line supervisor, it can't hurt to be reminded of the key work processes to manage safety performance that are entrusted to you:

- Setting and communicating work standards
- Teaching the right ways to do the work
- Determining who is qualified to perform the work
- Observing followers as they work
- Providing performance feedback—positive and corrective
- Rolling out safety policies and procedures
- Managing safety suggestions
- Running safety meetings
- Dealing with injuries and near misses

Those processes are critical because their output ultimately determines the level of safety performance the organization achieves. No leader has more organization power—the ability to make things happen the way you want them to—than do you. Your power comes in the form of control and influence: control over the "things" that cause harm, and influence over the people who might be harmed.

No leader has higher Performance Visibility than you. If conditions and problems need to be raised to your leader, that's the time to Upward Lead. If you think that will be a tough conversation, break out The Case for Safety and invoke the Principle of Honest Dialog. They will always help you do the right thing.

CHAPTER 22

MAKING A DIFFERENCE

> It's easy to make a buck. It's much tougher to make a difference.
>
> —*Tom Brokaw*

You've reached the end of the book: now it's time for you to make a decision.

Alive and Well began with a simple observation: because no follower wants to get hurt, and no leader wants to see any follower being hurt, managing safety should be easy. But in the real-life practice of leadership, sending every follower home alive and well at the end of every day may well be the toughest undertaking by any leader. Standing in the way of that noble goal is a formidable set of safety leadership challenges. Dealing with *all* those challenges—it really is all, as a leader doesn't get to pick and choose which ones to take on and which to ignore—is the job of the leader. Managing safety performance is the most important thing every supervisor, manager, and executive has to get done.

Followers are the biggest beneficiaries of safety performance—they go home alive and well at the end of the day—but it falls to their leaders to make that happen.

What you have learned in the first 21 chapters are answers to the questions you need to lead and manage safety performance.

- *What is leadership?*

The Leadership Model shows you the answer. Followers make you their leader. As their leader, you return the compliment by leading through words and actions. In the opinion of your followers—which is the only opinion that matters—your actions speak louder than your words.

- *When do I lead?*

Once your followers see you as their leader, you will always be looked as their leader. No leader ever has an off-camera moment. But there are certain times and places when your followers pay particular attention to what you do and say: those are the **Moments of High Influence**.

Followers determine the Moments: it's their choice as to when they pay close and careful attention, and when they tune you out, paying no real attention.

Alive and Well at the End of the Day: The Supervisor's Guide to Managing Safety in Operations, Second Edition. Paul D. Balmert.
© 2023 John Wiley & Sons, Inc. Published 2023 by John Wiley & Sons, Inc.

But there are a predictable set of causes that give rise to your Moments. They can be caused by the follower, by the leader, by an event and by a procedure.

The best time for you to lead is when your followers are paying attention. When your followers are paying attention, by definition what you do is leading.

- *How do I lead?*

The long list of what leaders do to lead reduces down to "use words" and "take actions." Taking action is always more impactful on your followers because it represents something they can see. But there is a useful and important set of functions performed by your words: they can explain, elicit, excite, and engage.

That said, the influence of your words as a leader is significantly affected by body language and credibility, both of which are traceable to your actions.

- *Why must safety always come first?*

That's the **Case for Safety**, summed up in three simple questions: What are the really important things in your life? How would a serious injury affect all those important things in your life? Is there any follower of yours whose answers to those two questions would be any different than yours?

Once you understand the **Case for Safety**, you have all the motivation you need to make a difference—and you won't settle for anything less. A refinery manager said it best: "I'm not going to stop until we get to zero. And then I'm not going to stop."

SAFETY LEADERSHIP TOOLS

In broad strokes, the answer to the question "How do I lead?" is simply to use words and take actions. That's something every leader already knows. What's been explained and examined in this book are the specific practices to do that, and do that well: these are what are called best practices. What you've found in the pages isn't just what to do; it's how to do that well, how it works, and why to do it that way. As a result, what you now have isn't just knowledge; it is understanding.

The answer to the question of what to do to lead is found in leadership practices. You've learned these, and a different type of tool, not used so much to lead followers as to aid and guide you as a leader. These tools help you make sense of what's going on, create order out of what seems disorder, understand your power, and keep you on the right path.

The best example of this type of leadership tool is the Case for Safety: it serves as your compass. You don't need a compass when the weather is perfect, and can see exactly where you're heading. Every leader knows how important safety is; the problem is, in the heat of the battle, when there's a production crisis, a threatening customer, or an unhappy boss, it's easy to lose sight of what matters most. That's the time to get out your compass and make sure you're headed in the right direction.

By way of a different analogy, Moments of High Influence serve as a wristwatch. What time is it? It's a Moment of High Influence: now is the perfect time for me to lead.

Control versus Influence gives you a formula to understand your power: as a leader you have control and you have influence; together they add up to your power. A leader is never powerless, and usually has a lot more power than they think. Understand that, and put it to good use.

The Injury Triangle gives you a model to understand who and what to mange: objects, energy and people. They're all important.

All of these combine to create a set of safety leadership and management tools. Before you picked up the book, you had your own set of tools; now you have many more. The more tools you have, the more options you have, and the better you will be at leading.

As a highly successful industrial leader, Van Long, reminds all leaders, "Leadership commitment without capability won't produce excellence."

LEADERS MUST LEAD

Interested in why a few mediocre companies morphed into superb performers, business professor and author Jim Collins launched a research project to determine what made the difference. Was it product innovation? Sales and marketing? Information technology? Simple random variation? Or something else?

What Collins learned by comparing the best with the rest was both remarkable and stunningly simple: leadership was the difference that made the difference in business performance. His best-selling book, *Good to Great*, details his research process, comparative performance data, and this remarkable finding.

But by demonstrating that leaders really are the difference, Collins also succeeded in proving, one more time, there really is nothing new under the Sun. The first book on managing a business, written in 1954, said exactly the same thing. The author of *The Practice of Management*, Peter Drucker, wrote, "Leadership is of utmost importance. Indeed, there is no substitute for it."

As Drucker saw it, "The purpose of an organization is making common men do uncommon things."

That is exactly what great safety performance is. In the safest operations, as people go about doing their work, they do things in a way that is anything but commonplace. That's what makes them the safest. What they do better—differently—is the product of leadership—by *their* leaders.

You might be thinking, "But I'm just one leader, and my operation doesn't have a great safety culture." Don't make the mistake of thinking a leader is a creature of the company safety culture.

It you think that way, it'll lead you to conclude, "When the culture finally changes, so will the safety performance of my followers." That completely discounts the fact that there are numerous examples of great safety performance to be found within otherwise mediocre organizations: a division, site, project, department, shift, or crew that far outperforms their peers. When you find that, you'll also find the fingerprints of a great leader all over the causes of that result.

It is as if the leader has created a counter-culture. It used to be said of the legendary football coach, Paul "Bear" Bryant, "He could take his team and beat yours; he could take your team and beat his."

There are leaders to be found who've proven perfectly capable of doing exactly that. Their safety leadership practices are the ones you've learned in *Alive and Well*.

TOUGH CHALLENGES

So, what are you going to do about all those tough safety challenges you face—every day?

Taking them on may well be the toughest assignment you have as a leader. These challenges demand time: your precious time as the leader, and the time it can take for the effect of your leadership to show up at the bottom line of safety performance. That time can seem like an eternity.

Because the challenges in one way or another ultimately reduce down to human behavior, their root cause can't really be "fixed" once and for all. Instead they require ongoing management. The work is unrelenting: no sooner will one challenge go into remission than another one will step to the forefront. Get the new people fully trained, they can become complacent just like their senior peers.

Managing safety performance often seems a thankless task. When you're successful, rarely will your followers take the time to thank you for keeping them safe. More likely they will complain about all the things you make them do to work safely. If you look up to your leaders for a sympathetic ear, you may be disappointed. Success isn't so much noticed as what's expected. When safety performance is good, your leader wants to talk about the business, not non-existent safety problems.

On the other hand, when things don't go according to plan and injuries and incidents happen, you can feel as though you're powerless to change the direction and improve the level of performance. Safety performance seems immune to any form of management or leadership.

That feeling isn't limited to front-line leaders: some of the most powerful executives on the planet have been incapable of managing safety performance successfully. As one once put it, "I feel like I'm leading, and nobody's following."

When things don't go well, it feels as though the real power in an organization lies somewhere else: with leaders up the chain of command, with followers on the job, bound up in the culture, a function of random events. It can leave a leader enormously frustrated: "What's the use? No matter what I do, it's not making any difference."

Making a difference is the whole point of leadership—and *Alive and Well*.

Leadership may seem like the invisible hand that moves followers to work safely, but how that hand of leadership functions can be discovered and understood by observation and learned by role modeling. Through interviews, Collins found the core leadership practices of top executives who successfully led companies from good to great performance. In a similar vein, from observation and experience, *Alive and Well* has defined and described the practices three generations of leaders have used to successfully lead and manage safety performance. The leadership practices described here are the ones that have been employed by leaders spanning the organization, from the boardroom to the shop floor that have proven to make a meaningful difference in performance.

For the most part, these practices are simple things leaders actually do to lead and create motivated followers. Peter Drucker described those kind of practices as "humdrum...requiring no genius, only application...things to do rather than talk about." That should be seen as good news for every leader. The leadership practices that make the difference in sending people home safe perfectly fit Drucker's characterization: they're found in the everyday details of running the business, and running it safely. At first glance those leadership practices might appear humdrum, but don't be fooled: those humdrum activities are the backbone of leadership.

It's been seven decades since Drucker wrote *The Practice of Management*, but the most fundamental truth about successfully managing a business hasn't changed: there is no substitute for leadership.

Not that this truth has stopped many organizations and their leaders from trying all kinds of alternatives to better leadership as a means to achieve those uncommon results, in business and in safety. It's easy to understand the temptation to jump on the latest innovation: launching a new safety program or process is easier—and more glamorous—than the tough duty of focusing on driving execution by leading better. The reality is that these alternatives often come across to the followers in the outfit as "the flavor of the month." In part because they are. Sure, there's a proper place for those approaches, but there is simply no substitute for leadership.

Programs and processes come and go like fads. Leadership is an entirely different matter: the best leadership practices stand the test of time.

IT'S YOUR CHOICE

Now we're to the moment of truth. Time for you to make a decision: what are you going to do with what you've learned from reading *Alive and Well*?

Your choice is simple, but not easy. You can learn. Or you can learn *and* put what you learned into practice. Do the former, you will have succeeded in acquiring knowledge, but nothing more. That makes your time spent reading the book an academic exercise. If you're a front-line supervisor and you make that choice, here's what your practice of safety leadership looks like:

- Run mediocre safety meetings
- Tell followers: "Put your PPE on!"
- Ignore small problems
- Trust "communicating" the new policy will make change happen
- Leave it up to your followers to get advice and philosophy from your boss or their peers
- Enter your safety data into the information system

If you're a manager:

- Forego coaching up supervisors who report to you
- Run mediocre safety meetings when it's your turn
- Manage from the office
- Depend on the current safety metrics to reveal what's really going on
- Hope your Performance Visibility is sufficient

If you're an executive:

- Rely on current safety metrics to provide Performance Visibility
- See safety execution as the job of those at the front line in operations
- Want great safety performance without ensuring leaders posses capability to achieve that level of performance
- Sign off on next big thing instead of focusing on execution

The alternative is to put the practices and tools found in *Alive and Well* into practice. Well executed, here's what that choice would look like:

If you're a front-line supervisor:

- Safety meetings are a good use of time
- Behavior correction has a long-lasting effect
- MBWA and understanding problems increase Performance Visibility
- Holding followers accountable solves a lot of individual performance problems
- Giving positive feedback and asking Darn Good Questions makes the job of being the front-line leader a lot easier.
- Your Early Warning Indicators alert you to problems before—not after

If you're a manager:

- Your safety meetings become a better use of time
- You have a model to evaluate and coach up your front-line supervisors
- MBWA and understanding small problems increase your Performance Visibility
- Better safety measures give you a sense of the actual level of safety performance
- More small problems surface and are dealt with before they become serious
- Managing accountability improves individual performance

If you're an executive:

- You successfully model specific leadership practices that you want to see followed down the chain of command
- Better safety measurement give you a sense as to the actual level of safety performance
- Your Performance Visibility increases
- Organization culture moves in the direction you want it to

Whether you put the tools to good use or put the book up on the shelf, they're both a decision. A decision that you control.

MAKING A DIFFERENCE

Putting this knowledge and practices to good use won't guarantee safety performance. But you can rest assured you have done everything in your power as a leader to manage safety performance as well as you possibly can. As a front-line supervisor, that is all that you can do.

But if you're a leader of other leaders, there's something else you can do: you can raise the leadership skills of the leaders who are your followers. That, too, is part of your job.

Do that and you will make a difference. The world in which you work and lead will be a better place, and you will have done your duty to see to it that your followers go home, alive and well at the end of the day.

As you now completely understand, that is the most important work of every leader.

REFERENCES

Bloomberg (2022). "Suncor replaces CEO Mark Little after oil sands mine death." https://www.bloomberg.com/news/articles/2022-07-08/suncor-replaces-ceo-mark-little-after-fatality-in-oil-sands-mine. Accessed 2022 July 8.

Bossidy, L. and Charan, R. (2002). *Execution: The discipline of getting things done.* New York: Crown Business.

Collins, J. (2001). *Good to great.* New York: Harper Collins.

Crosby, P. B. (1980). *Quality is free.* New York: New American Library.

Duhigg, C. (2012) *The power of habit.* New York: Random House.

Drucker, P. F. (1954). *The practice of management.* New York: Harper & Row.

DuBrul, R. (1992–unpublished). The development of effective peer management teams.

Hrebiniak, L. "Three reasons why good strategies fail: Execution, execution, execution." Knowledge@Wharton, August 10, 2005. http://knowledge.wharton.upenn.edu/article.cfm?articleid=1252. Accessed 2008 January 3.

Feynman, R. (1988). *What do you care what other people think? Further adventures of a curious character.* New York: W.W. Norton & Company.

Gladwell, M. (2002). *The tipping point.* Boston: Little, Brown and Company.

Grove, A. S. (1983). *High output management.* New York: Vintage Books.

Kletz, T. (1998). *What went wrong? Case studies of process plant disasters.* Elsevier Science & Technology Press.

Larkin, T. J. and Larkin, S. (1994). *Communicating change.* New York: McGraw-Hill.

Lowman, J. (1995). *Mastering the techniques of teaching.* New York: Wiley.

Mankins, M. and Steele, R. (2005). Turning great strategy into great performance. *Harvard Business Review*, July, 64–72.

NASA. (2003, August). Columbia accident investigation board report volume 1. http://anon.nasa-global.speedera.net/anon.nasa-global/CAIB/CAIB_lowres_full.pdf

McGehee, W. and Thayer, P. (1961). *Training and business and industry.* New York: Wiley.

OSHA (2019). Using leading indicators to improve safety and health outcomes. https://www.osha.gov/sites/default/files/OSHA_Leading_Indicators.pdf

Oshry, B. (1994). *In the middle.* Boston: Power & Systems.

Pelz, D. (1999). *Dave Pelz's short game bible.* New York: Broadway Books.

Peters, T. J. and Waterman, R. H. (1982). *In search of excellence.* HarperCollins.

Schwarzkopf, N. (1992). *It doesn't take a hero.* New York: Bantam Books.

Tampa Bay Times (2022). Teco pleads guilty, faces $500,000 fine in 2017 explosion that killed 5. https://www.tampabay.com/news/business/2022/05/06/teco-pleads-guilty-faces-500000-fine-in-2017-explosion-that-killed-5/. Accessed 2022 May 6.

The Sydney Morning Herald (2020). 'National tragedy': call for accountability over Rio Tinto destruction. June 5. https://www.smh.com.au/national/a-national-tragedy-activists-want-blood-for-rio-tinto-rock-shelter-blast-20200605-p54zzv.html

US Chemical Safety Board (2005). Investigation report refinery explosion and fire https://www.csb.gov/assets/1/20/csbfinalreportbp.pdf?13841

US Chemical Safety Board (1999). Summary report nitrogen asphyxiation. https://www.csb.gov/assets/1/20/final_union_carbide_report.pdf?13768

US Department of Justice (2021). Boeing charged with 737 Max fraud conspiracy and agrees to pay over $2.5 billion. Accessed 2022 January 7. https://www.justice.gov/opa/pr/boeing-charged-737-max-fraud-conspiracy-and-agrees-pay-over-25-billion

Vaughan, D. (1996). *The Challenger launch decision*. Chicago: University of Chicago Press.

Wall Street Journal (2010). Crew argued over drilling plan before rig explosion. https://www.wsj.com/articles/SB10001424052748704414504575244812908538510. Accessed 2022 May 15.

Ward K. (2006). The Charleston Gazette. Online Edition. May 4, 2006. I think I said "they're alive," rescuer tells hearing panel. http://wvgazette.com/News/TheSagoMineDisaster/200605040007. Accessed 2009 March 24.

INDEX

Alive and Well at the End of the Day: The Supervisor's Guide to Managing Safety in Operations,
Second Edition. Paul D. Balmert.
© 2023 John Wiley & Sons, Inc. Published 2023 by John Wiley & Sons, Inc.